热分析应用手册
Application Handbook
Thermal Analysis

热塑性聚合物
Thermoplastics

M.Zouheir Jandali　Georg Widmann　著

陆立明　唐远旺　蔡艺　译

本应用手册提供精选的应用实例。实验由瑞士梅特勒－托利多热分析实验室采用在每个应用中描述的特定仪器完成的,过程严谨,并以最新知识为依据对实验结果进行计算。
然而,这并非意味着读者无需用自己的适合样品的方法、仪器和用途进行亲自测试。由于对实例的效仿和应用是无法控制的,所以我们当然无法承担任何责任。
使用化学品、溶剂和气体时,必须遵循常规安全规范和制造商或供应商给予的使用指南。
This application handbook presents selected application examples. The experiments were conducted with the utmost care using the instruments specified in the description of each application at METTLER TOLEDO Themal Analysis Lab in Switzerland. The results have been evaluated according to the current state of our knowledge.
This does not however absolve you from personally testing the suitability of the examples for your own methods, instruments and purposes. Since the transfer and use of an application is beyond our control, we cannot of course accept any responsibility.
When chemicals, solvents and gases are used, general safety rules and the instructions given by the manufacturer or supplier must be observed.

出版前言

本书是《热分析应用手册》的《热塑性聚合物》分册。

本套《热分析应用手册》，按行业及专门技术应用，计划出版若干分册，分别有：

- 热分析基础
- 热塑性聚合物
- 热固性聚合物
- 弹性体
- 食品
- 药物
- 无机物
- 化学品
- 热重-逸出气体分析

本套书既注重实用性，又注重学术性。它们不仅可以作为应用手册查询，也可以作为实验指南，如选择合适的测试技术和方法、制备和处理样品、设定实验参数等。手册中的所有应用实例都经过认真挑选，实验方法精心设计，测试曲线重复可靠，数据处理严格谨慎，对实验结果的解释和对实验结论的推导科学合理。每一个应用，都是一篇短小的学术论文。这样编写的热分析应用手册可谓独具匠心。

本套手册的作者，都是长期从事热分析技术应用研究的学者和工程师，具有丰富的经验和深厚的造诣。

本套手册面向所有用到热分析和对热分析感兴趣的科学家、工程师、学生（特别是研究生）及其他科技工作者，适合所有热分析仪器的直接使用者。

本套手册也是很好的教学参考书。

本套书以中英文对照方式出版，读者可以阅读中文，同时可对照原著。无论对热分析工作者，还是热分析学习者，应该都有帮助和裨益。

译文甚至原著中，有错误之处，望读者指正，以使能在再版时改正，不胜感谢。

陆立明

2008年6月，上海

目 录

应用列表	III
1. 热分析导论 Introduction to Thermal Analysis	1
1.1 差示扫描量热法(DSC) Differential Scanning Calorimetry	1
1.1.1 常规 DSC Conventional DSC	1
1.1.2 温度调制 DSC Temperature-modulated DSC	2
1.1.2.1 ADSC	2
1.1.2.2 IsoStep	4
1.1.2.3 TOPEM™	5
1.2 热重分析(TGA) Thermogravimetric Analysis	5
1.3 热机械分析(TMA) Thermomechanical Analysis	7
1.4 动态热机械分析(DMA) Dynamic Mechanical Analysis	8
1.5 与 TGA 的同步测量 Simultaneous Measurements with TGA	11
1.5.1 同步 DSC 和差热分析(DTA,SDTA) Simultaneous DSC and Differential Thermal Analysis	11
1.5.2 析出气体分析(EGA) Evolved Gas Analysis	12
1.5.2.1 TGA-MS	12
1.5.2.2 TGA-FTIR	13
2. 聚合物的结构和性能 Structure and Behavior of Polymers	15
2.1 聚合物领域的一些定义 Some Definitions in the Field of Polymers	15
2.2 聚合物的物理结构 Physical Structure of Polymers	16
2.3 热塑性聚合物 Thermoplastic Polymers	18
2.3.1 无定形塑料 Amorphous Plastics	18
2.3.2 半结晶塑料 Semicrystalline Plastics	19
3. 热塑性聚合物的重要领域 Important Fields of Thermoplastic Polymers	21
4. 热塑性聚合物的应用一览表 Application Overview of Thermoplastic Polymers	23
5. 热塑性聚合物的特征温度表 Table of characteristic temperatures of thermoplastic polymers	24
6. 重要热塑性聚合物的性能和典型的热分析应用 Properties of Important Thermoplastic Polymers and Typical TA Applications	26
6.1 聚乙烯,PE Polyethylene	26
6.2 乙烯/醋酸乙烯共聚物,E/VAC Ethylene/Vinylacetate Copolymer	26
6.3 聚丙炳,PP Polypropylene	27
6.4 聚苯乙烯,PS Polystyrene	27
6.5 聚氯乙烯,PVC Polyvinyl Chloride	28
6.6 聚醋酸乙烯,PVAC Polyvinyl Acetate	29
6.7 聚酰胺,PA Polyamide	29

6.8 聚对苯二甲酸乙二醇酯,PET Polyethylene Terephthalate ……………………………… 30
6.9 聚碳酸酯,PC Polycarbonate ……………………………………………………………… 30
6.10 聚甲醛,POM Polyoxymethylene ………………………………………………………… 31
6.11 聚四氟乙烯,PTFE Polytetrafluoroethylene …………………………………………… 31

7. 热塑性聚合物的应用 Applications of Thermoplastic Polymers …………………… 33
7.1 聚乙烯测试 Measurements on Polyethylene ………………………………………… 33
7.2 聚丙烯测试 Measurements on Polypropylene Based Material …………………… 58
7.3 聚苯乙烯的玻璃化转变 Glass Transition of Polystyrene ………………………… 67
7.4 聚氯乙烯的热分析测试 TA Measurements on Polyvinyl Chloride ……………… 72
7.5 聚酰胺及其共混物 Polyamides and Their Blends ………………………………… 82
7.6 聚对苯二甲酸乙二醇酯的热行为 Thermal Behavior of Polyethylene Terephthalate ……… 104
7.7 其它聚合物测试 Measurements on Other Polymers ……………………………… 115
7.8 热塑性弹体 Thermoplastic Elastomers ……………………………………………… 132
7.9 聚合物共混物和共聚物 Polymer Blends and Copolymers ………………………… 136
7.10 热塑性塑料及其产品的进一步测试 Further Measurements of Thermoplastics and Their Products …………………………………………………………………… 148

文献 Literature ……………………………………………………………………………… 159

应用列表 Application list

标题 Title	主题 Topics						方法 Methods				页码 Page
	玻璃化转变 Glass transition	结晶度/结晶 Crystallinity / crystallization	熔融 Melting	反应 Reaction	成分/含量 Composition / content	数据计算/实验 Evaluation / experimental	其他 Others	DSC / ADSC / IsoStep	TGA / TGA-EGA	TMA / DLTMA	DMA	
用峰温表征聚乙烯 PE, Characterization by Peak Temperature			●			●		●				33
用结晶度表征聚乙烯 PE, Characterization by Crystallinity		●	●					●				35
用转化率曲线表征高密度聚乙烯 PE-HD, Characterization by Conversion Curves			●			●		●				37
用结晶行为表征高密度聚乙烯 PE-HD, Characterization by Crystallization Behavior		●						●				39
来自不同制造商的高密度聚乙烯 PE-HD from Different Manufacturers		●	●					●				41
聚乙烯的熔融曲线和热历史 PE, Melting Curve and Thermal History			●			●	●					43
高密度聚乙烯电缆管线和再生料的鉴定 PE-HD, Identification of Cable Tubing and Recycled Material		●						●				45
高密度聚乙烯再生板的DSC DSC of Recycled sheets, Said to Be PE-HD		●	●				●					47
低密度聚乙烯的两个产品的比较 PE-LD, Comparison of Two Products			●				●	●				48
聚乙烯的氧化稳定性 PE, Oxidation Stability				●				●				50
用动态负载TMA测试交联聚乙烯 Cross-linked PE by Dynamic Load TMA			●			●				●		52
聚乙烯的TGA成分分析 PE, Compositional Analysis by TGA					●	●			●			54
高取向高密度聚乙烯纤维的DSC DSC of Highly Oriented PE-HD fiber		●	●					●				56
样品质量对聚丙烯的影响 PP, Influence of The Sample Mass		●	●			●		●				58
来自不同制造商的聚丙烯 PP from Different Manufacturers								●				60
空气中聚丙烯的DSC测试 PP, DSC Measurements in Air		●	●	●								62
空气中聚丙烯重复性 PP, Repeated Cycling in Air		●	●									64
聚丙烯:新料与再生料 PP, New Versus Recycled Material			●				●					66
聚苯乙烯的DSC曲线 DSC Curves of PS	●							●				67
用DLTMA测定聚苯乙烯的玻璃化转变 PS, Glass Transition By DLTMA	●					●				●		69
用DSC和TGA测试聚氯乙烯 PVC Measured by DSC and TGA	●			●		●		●	●			72

(续表)

标题 Title	主题 Topics							方法 Methods				页码 Page
	玻璃化转变 Glass transition	结晶度/结晶 Crystallinity / crystallization	熔融 Melting	反应 Reaction	成分/含量 Composition / content	数据计算/实验 Evaluation / experimental	其他 Others	DSC / ADSC / IsoStep	TGA / TGA-EGA	TMA / DLTMA	DMA	
未增塑聚氯乙烯的热稳定性 PVC-U, Thermal Stability				•		•			•			74
聚氯乙烯的 TMA 曲线与所加负载的关系 PVC, TMA Curves as A Function of Applied Load	•									•		76
聚氯乙烯和氯化聚氯乙烯的玻璃化转变 Glass Transition of PVC and Chlorinated PVC	•					•	•					78
聚氯乙烯增塑剂混合物的凝胶化 Gelation of A PVC Plasticizer Mixture	•					•					•	80
聚酰胺 6 的熔融行为 Polyamide 6, Melting Behavior		•	•				•					82
聚酰胺 6：新料与再生料 PA6, New Versus Recycled Material			•				•					84
玻璃纤维含量的测定 Determination of Glass Fiber Content					•				•			86
不同质量的聚酰胺 66 PA66, Different Qualities			•									88
用 TGA 和 DSC 测定聚酰胺 66 的水含量 Determination of The Moisture Content of PA66 by TGA and DSC					•	•	•	•	•			90
不同加工批次的聚酰胺 66/聚酰胺 6 PA66/PA6 Batches of Different Processability			•			•	•					92
错误认定的聚酰胺 6 和聚酰胺 66 PA6 and PA66, Mistaken Identity			•				•					94
聚酰胺 6/聚酰胺 66 共混物 PA6/PA66 Blend			•		•	•	•					96
聚酰胺 6 共混物 Polyamide 6 Blends			•				•					98
用 IsoStep DSC 测定聚酰胺 6 的玻璃化转变和水分含量 Glass Transition and Moisture Content of PA6 by IsoStep DSC	•											100
聚对苯二甲酸乙二醇酯的热历史 PET, Thermal History	•	•	•			•						104
聚对苯二甲酸乙二醇酯的焓松弛 PET, Enthalpy Relaxation	•					•						107
用动态负载 TMA 测定聚对苯二甲酸乙二醇酯的冷结晶 PET, Cold Crystallization by Dynamic Load TMA	•	•				•				•		110
聚对苯二甲酸乙二醇酯的动态热机械分析 Dynamic Mechanical Analysis of PET	•	•	•			•					•	112
聚甲基丙烯酸甲酯的玻璃化转变 PMMA, Glass Transition	•					•						115
聚甲醛的 DSC 测试 DSC Measurement of POM		•				•						117
聚乙二酸丙二醇酯的 DSC 测试 DSC Measurements of PPA	•	•	•			•						119

(续表)

标题 Title	主题 Topics							方法 Methods				页码 Page
	玻璃化转变 Glass transition	结晶度/结晶 Crystallinity / crystallization	熔融 Melting	反应 Reaction	成分/含量 Composition / content	数据计算/实验 Evaluation / experimental	其他 Others	DSC / ADSC / IsoStep	TGA / TGA-EGA	TMA / DLTMA	DMA	
高温聚合物 High Temperature Polymers	•		•					•				121
用 DSC 和 TMA 测定聚四氟乙烯多晶态 PTFE Polymorphism by DSC and TMA						•		•		•		123
用 DMA 和 DSC 表征聚四氟乙烯 Characterization of PTFE by DMA and DSC	•					•		•			•	125
用 ADSC 测定聚醚酰亚胺的玻璃化转变 PEI, Glass Transition by ADSC	•					•		•				128
聚醚酰亚胺的 DMA 分析 DMA Analysis of PEI	•					•					•	130
酯类热塑性弹性体 TPE-E, Ester-based Thermoplastic elastomer			•			•		•				132
烯烃类热塑性弹性体 TPE-O, Olefin-based Thermoplastic Elastomer			•					•				134
用 DSC 测试聚碳酸酯和聚碳酸酯/ABS 共混物 PC and a PC/ABS Blend Measured by DSC	•							•				136
用 DSC 和 TMA 表征乙烯/醋酸乙烯共聚物 E/VAC, Characterization by DSC and TMA	•		•			•		•		•		138
用 DSC 测定丙烯腈/丁二烯/苯乙烯共聚物的玻璃化转变 ABS Glass Transition by DSC	•		•		•			•				140
甲基丙烯酸甲酯/丁二烯/苯乙烯共聚物的 DSC 和 DMA 测试 DSC and DMA Measurements of An MBS Copolymer	•	•				•		•			•	142
聚 ε-己内酰胺/聚四氢呋喃共聚物的结晶和熔融 Crystallization and Melting of PCL/PTHF Copolymers			•	•		•		•			•	145
聚丙烯/聚乙烯共聚物的定性检查 PP/PE, Copolymer Identity Check			•					•				148
聚醋酸乙烯的玻璃化转变温度和增塑剂含量 PVAC, Glass Transition Temperature and Plasticizer Content	•				•	•	•					150
聚对苯二甲酸丁二醇酯共混物上涂膜的粘着性 Adherence of Paint on a PBT Blend			•		•	•						152
TMA 测试合成纤维 TMA Measurements on Synthetic Fibers						•				•		154
用 DMA 和 DSC 分析墨粉 Analysis of Toner Powder by DMA and DSC	•							•			•	156

1. 热分析导论 Introduction to Thermal Analysis

热分析是测试材料的物理和化学性能与温度的函数关系的一类技术的总称。在所有方法中,样品受控于加热、冷却或恒温温度程序。

测试可在不同气氛中进行,通常使用惰性气氛(氮气、氩气、氦气)或氧化气氛(空气、氧气)。某些情况下,在测试中气体从一种气氛切换到另一种气氛。有时另一个可选择变化的参数是气体压力。

DSC 还可与能同步观察样品的仪器联用(DSC 显微镜方法)或用不同波长的光照射(光量热法)。

Thermal analysis is the name given to a group of techniques used to measure the physical and chemical properties of materials as a function of temperature. In all these methods, the sample is subjected to a heating, cooling or isothermal temperature program.

The measurements can be performed in different atmospheres. Usually either an inert atmosphere (nitrogen, argon, helium) or an oxidative atmosphere (air, oxygen) is used. In some cases, the gases are switched from one atmosphere to another during the measurement. Another parameter sometimes selectively varied is the gas pressure.

DSC can also be used in combination with instruments that allow the sample to be simultaneously observed (DSC microscopy) or exposed to light of different wavelengths (photocalorimetry).

1.1 差示扫描量热法(DSC) Differential Scanning Calorimetry

在 DSC 中,测量样品吸收和放出的热量。DSC 可用于研究物理转变(玻璃化转变、结晶、熔融和挥发成分的蒸发)和化学反应这样的热效应,所获得的信息表征样品的热性能和组成。此外,还可测定热容、玻璃化转变温度、熔融温度、反应热和反应程度这样的性能。

In DSC, the heat flow to and from the sample is measured. DSC can be used to investigate thermal events such as physical transitions (the glass transition, crystallization, melting, and the vaporization of volatile compounds) and chemical reactions. The information obtained characterizes the sample with regard to its thermal behavior and composition. In addition, properties such as the heat capacity, glass transition temperature, melting temperature, heat and extent of reaction can also be determined.

1.1.1 常规 DSC Conventional DSC

常规 DSC 采用线性温度程序,样品和参比物(或只是空坩埚)以线性速率加热或冷却,或在某些情况下保持在恒定温度(即恒温)。经常几部分程序即所谓的程序段连接在一起生成一个完整的温度程序。聚合物的典型 DSC 曲线如图 1.1 所示。

测试开始时曲线上的变化是由于初始的"启动偏移"(1)。在该瞬变区域,状态突然从恒温模式变为线性升温模式。启动偏移的大小取决于样品热容和升温速率。在玻璃化转

Conventional DSC employs a linear temperature program. The sample and reference material (or just an empty crucible) are heated or cooled at a linear rate, or in some cases, held at a constant temperature (i. e. isothermally). Often several partial programs or so-called segments are joined together to form a complete temperature program. A typical DSC curve of a polymer is shown schematically in Figure 1.1.

The change in the curve at the beginning of the measurement is due to the initial "startup deflection" (1). In this transient region, the conditions suddenly change from an isothermal mode to a linear heating mode. The magnitude of the startup deflection depends on the heat capacity of the sample and the heating rate.

变区(2)，样品的热容增加，因而可观察到一个吸热台阶。冷结晶(3)发生在玻璃化转变以上。结晶容易的聚合物被加热至熔点以上，然后骤冷以遏制微晶的形成。这样的聚合物在玻璃化转变之上重结晶。继续加热时，发生熔融(4)。在较高的温度开始分解(6)。在熔融和分解之间有些物质可能汽化(5)。

实验中使用的保护气氛的种类经常对涉及的反应有影响。

At a glass transition (2), the heat capacity of the sample increases and therefore an endothermic step is observed. Cold crystallization(3) occurs above the glass transition. Polymers that crystallize readily are heated to above the melting point and quench-cooled to suppress the formation of crystallites. Such polymers recrystallize above the glass transition. On further heating, melting (4) takes place. At higher temperatures, decomposition (6) begins. Some substances may vaporize(5) between the melting and decomposition.

The type of purge gas used in the experiment often has an influence on the reactions involved.

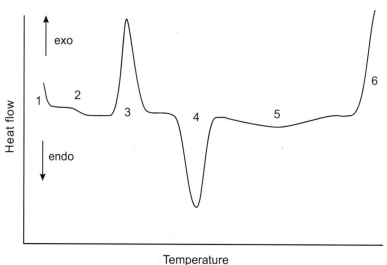

图 1.1　图示聚合物 DSC 曲线：1 初始启动偏移；2 玻璃化转变；3 冷结晶；4 熔融；5 汽化；6 分解
Fig. 1.1　Schematic DSC curve of a polymer：1 initial startup deflection；2 glass transition；3 cold crystallization；4 melting；5 vaporization；6 decomposition

转变和反应可通过冷却样品和再次测试它来区分——化学反应是不可逆的，而熔化了的结晶材料当冷却或二次加热时会重新结晶。玻璃化转变也是可逆的，但经常在玻璃化转变的第一次加热测试中观察到的焓松弛是不可逆的。

Transitions and reactions can be differentiated by cooling the sample and measuring it again-chemical reactions are irreversible whereas crystalline materials melt then crystallize again on cooling or on heating a second time. Glass transitions are also reversible but not the enthalpy relaxation often observed in the first heating measurement of a glass transition.

1.1.2　温度调制 DSC　Temperature-modulated DSC

1.1.2.1　ADSC

调制 DSC(ADSC)是一种特别类型的温度调制 DSC(TMDSC)。与常规 DSC 不同，小周期温度变化叠加在线性温度程序上。温度程序的特

Alternating DSC (ADSC) is a particular type of temperature-modulated DSC (TMDSC). In contrast to conventional DSC, the linear temperature program is overlaid with a small periodic temperature change. The temperature program is characterized

征为基础加热速率、温度振幅和周期性变化温度的持续时间(图1.2)。采用准恒温测试,基础加热速率也可为零。

by the underlying heating rate, the temperature amplitude and the duration of the periodically changing temperature (Fig. 1.2). With quasi-isothermal measurements, the underlying heating rate can also be zero.

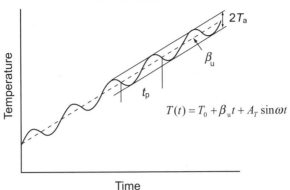

图 1.2 典型 ADSC 温度程序:β_U 为基础加热速率,A_T 为温度振幅,t_P 为周期,$2\pi/P$ 为角频率 ω,P 为正弦波的周期

Fig. 1.2 Typical ADSC temperature program: β_U is the underlying heating rate, A_T the temperature amplitude, t_P period, The angular frequency ω is defined as $2\pi/P$ where P denotes the period of the sine wave.

由于温度的调制,所测得的热流呈周期性变化。该热流能分离成两部分,如图 1.3 所示。信号平均生成基本信号(总热流),它相当于常规

As a result of temperature modulation, the measured heat flow changes periodically. This can be separated into two parts as shown in Figure 1.3. Signal averaging yields the underlying signal (total heat flow), which corresponds to the conventional DSC

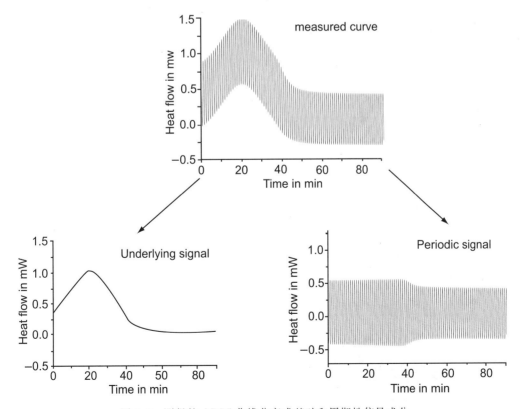

图 1.3 测得的 ADSC 曲线分离成基础和周期性信号成分

Fig. 1.3 Separation of the measured ADSC curve into the underlying and the periodic signal components

DSC曲线。作为附加信息,还得到周期性信号成分。可逆热流为能够直接跟上加热速率变化的热流成分,从同相比热计算得到。总热流与可逆热流的差值得到不可逆热流。本技术的一个优势是能将同时发生的过程分开。例如,化学反应过程中的热容变化可直接测量。

ADSC曲线的数值计算基于傅立叶分析。复合比热 c_p^* 的模量用下面的等式计算:

curve. As additional information, one also obtains the periodic signal component. The reversing heat flow corresponds to the heat flow component that is able to follow the heating rate change directly and is computed from the in-phase heat capacity. The difference between the total heat flow and the reversing heat flow yields the non-reversing heat flow. One advantage of this technique is that it allows processes that occur simultaneously to be separated. For example, the change in heat capacity during a chemical reaction can be measured directly.

The evaluation of the ADSC curves is based on Fourier analysis. The modulus of the complex heat capacity c_p^* is calculated using the equation,

$$|C_p^*| = \frac{A_\Phi}{A_\beta} \cdot \frac{1}{m}$$

式中 A_Φ 和 A_β 为调制热流和加热速率的振幅,m 为样品质量。ADSC 热流信号与加热速率之间的相角用于计算同相 c_p。

where A_Φ and A_β denote the amplitudes of the modulated heat flow and heating rate, and m the sample mass. The phase angle between the ADSC heat flow signal and the heating rate is used to calculate the in-phase c_p.

1.1.2.2 IsoStep

IsoStep 是一种特殊类型的温度调制 DSC。在该方法中,温度程序由很多开始和结束为恒温段的动态程序段组成(图 1.4)。

IsoStep is a special type of temperature-modulated DSC. In this method, the temperature program consists of a number of dynamic segments that begin and end with an isothermal segment (Fig. 1.4).

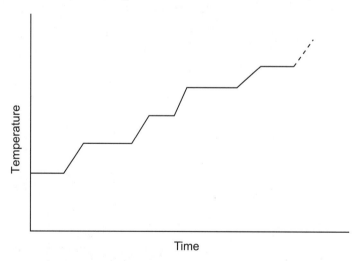

图 1.4 IsoStep 温度程序由不同的恒温和动态段组成
Fig. 1.4 IsoStep temperature program consisting of different isothermal and dynamic segments

恒温段能让动态段的恒温漂移获得修正,结果得到更好的热容准确性。恒温台阶还可能包含动力学信息,

The isothermal segments allow the isothermal drift of the dynamic segments to be corrected. This results in better heat capacity accuracy. The isothermal step may also contain kinetic

例如化学反应。比热可用蓝宝石参比样进行测定,而动力学效应能从热容变化中分离开。

information, for example of a chemical reaction. Heat capacity determinations can be made using a sapphire reference sample, and kinetic effects can be separated from changes in heat capacity.

1.1.2.3 TOPEM™

TOPEM™是高级温度调制DSC技术,基于对DSC(仪器和样品两者)对随机调制基础温度程序响应的全面数学分析(图1.5)。由于温度脉冲是随机分布的,系统在宽频范围而不是只在某单一频率(ADSC)内服从于温度振荡。振荡式输入信号(加热速率)和响应信号(热流)的相关分析能得到比常规温度调制DSC多得多的信息,不仅能将可逆与不可逆效应分开,而且还能测量样品的准稳态热容和测定频率依赖的热容值。这可用来在一次测试中就区分开频率依赖的松弛效应(例如玻璃化转变)和非频率依赖的效应(例如化学反应)。

TOPEM™ is an advanced temperature-modulated DSC technique that is based on the full mathematical analysis of the response of a DSC (both the apparatus and the sample) to a stochastically modulated underlying temperature program (Fig. 1.5). Due to the randomly distributed temperature pulses, the system is subjected to temperature oscillations over a wide frequency range and not just at one single frequency (ADSC). An analysis of the correlation of the oscillating input signal (heating rate) and the response signal (heat flow) provides much more information than can be obtained using conventional temperature-modulated DSC. Not only can reversing and non-reversing effects be separated, but the quasi-static heat capacity of the sample is also measured and frequency-dependent heat capacity values are determined. This can be used to distinguish between frequency-dependent relaxation effects (e. g. glass transitions) and frequency-independent effects (e. g. chemical reactions) in one single measurement.

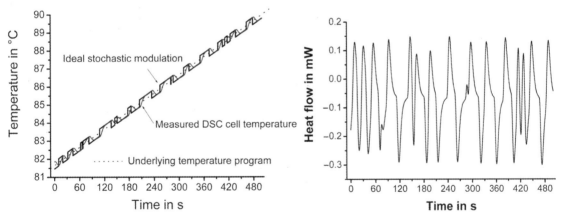

图 1.5 TOPEM™方法中炉体设定值的温度曲线(黑线),炉体温度(红线)产生在平均值上下变动的加热速率。如右图所示,流向样品的热流也不规则变动。

Fig. 1.5. Temperature curve of the furnace set value (black line) in a TOPEM™ method in which the furnace temperature (red curve) generates a heating rate that fluctuates around a mean value. The heat flow to the sample also fluctuates irregularly as shown in the diagram on the right.

1.2 热重分析(TGA)　Thermogravimetric Analysis

当样品被加热时,质量开始减少。失重可能产生于蒸发或有样品生成和逸出气体产物的化学反应。如果

When a sample is heated, it often begins to lose mass. This loss of mass can result from vaporization or from a chemical reaction in which gaseous products are formed and evolved from the

保护气氛不是惰性的,样品还能与气体反应。在某些情况下,样品质量也可能增加,例如在当生成产物是固体的氧化反应中。

在热重分析(TGA)中,测量样品质量的变化与温度或时间的函数关系。

TGA 提供关于样品性能及其成分的信息。如果样品分解产生于化学反应,则样品质量通常呈台阶状变化。台阶出现时的温度可表征该样品材料在所用气氛中的稳定性。

图1.6 所示为典型的 TGA 曲线。通过分析单独质量台阶的温度和高度能确定材料的成分。

sample. If the purge gas atmosphere is not inert, the sample can also react with the gas. In some cases, the sample mass may also increase, e. g. in an oxidation reaction if the product formed is a solid.

In thermogravimetric analysis (TGA), the change in mass of a sample is measured as a function of temperature or time.

TGA provides information on the properties of the sample and its composition. If the sample decomposes as a result of a chemical reaction, the mass of the sample often changes in a stepwise fashion. The temperature at which the step occurs characterizes the stability of the sample material in the atmosphere used.

Figure 1.6 shows a typical TGA curve. The composition of a material can be determined by analyzing the temper-atures and the heights of the individual mass steps.

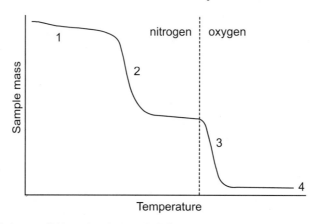

图1.6 图示 TGA 曲线:1 由于挥发性成分蒸发导致的失重;2 在惰性气氛中的热解;
3 当切换到氧化气氛后的碳黑燃烧;4 残留物。

Fig. 1.6 Schematic TGA curve: 1 loss of mass due to the vaporization of volatile components; 2 pyrolysis in an inert atmosphere; 3 combustion of carbon on switching from an inert to an oxidative atmosphere; 4 residue.

像水、残留溶剂或添加油这样的挥发性化合物在相对低的温度逸出。这样的化合物的排除取决于气体压力,在低压下(真空),相应的失重台阶移到低温,就是说,蒸发加速。分析在惰性气氛中的热解反应能确定含量(从台阶高度),甚至能确定材料的种类。

样品的碳黑或碳纤维含量可从切换到氧气气氛后的燃烧台阶的高度确定。残留填料、玻璃纤维和灰分由残留物台阶确定。测试曲线上由于浮力效应和气流速率而产生的小变

Volatile compounds such as water, residual solvents or added oils are evolved at relatively low temperatures. The elimination of such components depends on the gas pressure. At low pressures (vacuum), the corresponding mass loss step is shifted to lower temperatures, that is, vaporization is accelerated. The analysis of pyrolysis reactions in an inert atmosphere allows the content (from the step height) and possibly even the type of material to be determined.

The carbon black or carbon fiber content of a sample can be determined from the height of the combustion step after switching to an oxidative atmosphere. The residual filler, glass fiber or ash is determined from the residue. Small changes in the measurement curve due to buoyancy effects and gas flow rate can

化,可通过减去空白曲线得到修正。TGA 测试常常用 TGA 曲线的一阶微分(称为 DTG 曲线)显示。于是,TGA 曲线上由质量损失所致的台阶在 DTG 曲线上以峰形呈现。DTG 曲线相当于样品质量变化的速率。

分解台阶的温度范围在一定程度上受到气体产物扩散出样品的容易性的影响。当使用反应性气体时,样品表面气体交换的效率是关键。可以使用合适的坩埚(例如,30 μL 氧化铝坩埚这样的低壁高坩埚)和合适形状的样品(几个小颗粒或粉末)来降低测试时的扩散效应。

在 TGA 中,样品质量的变化被非常准确地测量。然而令人遗憾的是该技术不提供关于逸出气体分解产物性质的任何信息。不过用 TGA 与合适的气体分析器偶联(逸出气体分析 EGA),能分析这些产物。

be corrected by subtracting a blank curve. TGA measurements are often displayed as the first derivative of the TGA curve, the so-called DTG curve. Steps due to loss of mass in the TGA curve then appear as peaks in the DTG curve. The DTG curve corresponds to the rate of change of sample mass.

The temperature range of the decomposition steps is influenced to a certain extent by the ease with which the gaseous products are able to diffuse out of the sample. When reactive atmospheres are used, the efficiency of gas exchange at the surface of the sample is crucial. The effects of diffusion on the measurement can be reduced by using suitable crucibles (e.g. crucibles with low wall-heights such as the 30-μL alumina crucible) and by suitable sample geometry (several small pieces or powder).

In TGA, the change in mass of the sample is measured very accurately. Unfortunately, however, the technique does not provide any information about the nature of the gaseous decomposition products evolved. The products can however be analyzed by coupling the TGA to a suitable gas analyzer (evolved gas analysis, EGA).

1.3 热机械分析(TMA) Thermomechanical analysis

热机械分析测试样品在加热时的尺寸变化,在该技术中,连续测量带一定力且放置于样品表面的探头位置或位移与温度或时间的函数关系。图 1.7 所示为典型的 TMA 曲线。事实上探头施加的压力和样品的硬度决定了 TMA 实验是膨胀还是穿透测试。

在热膨胀测试中,探头在样品表面仅施加低压力。样品在整个相应的温度范围内被线性加热。线性热膨胀系数(CTE)直接从测试曲线计算。

在穿透实验中,探头施加大得多的压力。当对样品加热时,可直接测得样品的软化温度,材料在玻璃化温度处或熔融时软化。

Thermomechanical analysis measures the dimensional changes of a sample as it is heated. In this technique, the position or displacement of a probe resting on the surface of the sample with a certain force is continuously measured as a function of temperature or time. Figure 1.7 shows a typical TMA curve. The pressure exerted by the probe and the hardness of the sample determine whether the TMA experiment is in fact an expansion or a penetration measurement.

In the thermal expansion measurement, the probe exerts only a low pressure on the surface of the sample. The sample is heated linearly over the temperature range of interest. The linear coefficient of thermal expansion (CTE) is calculated directly from the measurement curve.

In a penetration experiment, the probe exerts a much greater pressure. The softening temperature can be directly measured when the sample is heated. Materials soften at the glass transition temperature or on melting.

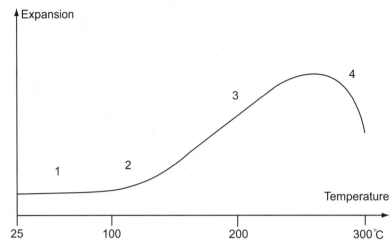

图 1.7 在低压缩应力下聚合物的图示 TMA 曲线：
1 低于玻璃化转变的膨胀;2 玻璃化转变(斜率变化);3 玻璃化转变以上的膨胀;4 塑性形变

Fig.1.7 Schematic TMA curve of a polymer under low compressive stress: 1 expansion below the glass transition; 2 glass transition (change of slope); 3 expansion above the glass transition; 4 plastic deformation

如果对样品施加周期性变化的力,样品尺寸也周期性变化。该测试模式称为动态负载 TMA,DLTMA,提供关于聚合物的粘弹行为的信息。从振幅和样品厚度能估算出样品的弹性模量(杨氏模量)。

使用弯曲附件,用此技术还能测量硬样品的弯曲行为。

可用专门的样品支架测量纤维和薄膜的尺寸变化。一个独特的应用是测量材料在溶剂中的溶胀行为。为此,样品放入一个小容器内,加入相应的溶剂,于是,在恒温下连续测量由于溶剂吸收导致的样品厚度随时间的变化。

If a periodically changing force is applied to the sample, the sample dimensions also change periodically. This measurement mode is called dynamic load TMA, DLTMA, and provides information on the viscoelastic behavior of polymers. The elastic modulus (Young's modulus) of the sample can be estimated from the amplitude and the sample thickness.

It is also possible to measure the bending behavior of hard samples with this technique using a bending accessory.

Special sample holders are available that allow the dimensional changes of fibers and films to be measured. A particular application is the measurement of the swelling behavior of materials in solvents. To do this, a sample is placed in a small container and the solvent of interest is added. The change in thickness of the sample due to solvent absorption is then contin- uously measured isothermally as a function of time.

1.4　动态热机械分析(DMA)　Dynamic Mechanical Analysis

在动态热机械分析中,测定动态模量与温度、频率和振幅之间的函数关系。

施加于样品的周期性(通常为正弦)变化的力在样品中产生周期性的应力。样品对该应力作出反应,仪器测量相应的形变行为,由应力和形变测定机械模量 M。无论测量剪

In dynamic mechanical analysis, a mechanical modulus is determined as a function of temperature, frequency and amplitude.

A periodically changing force (usually sinusoidal) appliedto the sample creates a periodic stress in the sample. The sample reacts to this stress and the instrument measures the corresponding deformation behavior. The mechanical modulus, M, is determined from the stress and deformation. Depending on the

切模量 G（施加剪切应力）还是杨氏模量 E（拉伸或弯曲），均取决于所加应力的类型。

样品不总是对周期性变化的应力作出瞬间响应——依赖于样品的粘弹性而发生一定时间的滞后。这是产生施加应力和形变之间相位移的原因。将相位移考虑进去，动态测得的模量用实数部分 M′和虚数部分 M″来描述。实数部分（储能模量）描述与周期性应力同相的样品响应，它是样品（可逆的）弹性的量度。虚数部分（损耗模量）描述相位移为 90°的响应部分，它是转化为热（因而不可逆地损失了）的机械能量的量度。相位移的正切 tan δ 还被称作损耗因子，是材料阻尼性能的量度。模量和 tan δ 依赖于温度和测试频率。在室温下，橡胶的典型储能模量在 0.1MPa 至 10MPa 之间。

图 1.8 中的曲线显示了无定形和半结晶聚合物的储能和损耗模量与温度函数关系的典型行为。

type of stress applied, either the shear modulus, G (with shear stress) or the Young's modulus, E (with stretching or bending) is measured.

The sample does not always immediately react to the periodically changing stress-a certain time delay occurs that depends on the viscoelastic properties of the sample. This is the cause of the phase shift between the applied stress and the deformation. To take this phase shift into account, the dynamically measured modulus is described by a real part M′ and an imaginary part M″. The real part (storage modulus) describes the response of the sample in phase with the periodic stress. It is a measure of the (reversible) elasticity of the sample. The imaginary part (loss modulus) describes the component of the response that is phase-shifted by 90°. This is a measure of mechanical energy converted to heat (and therefore irreversibly lost). The tangent of the phase shift, tan δ, is also known as the loss factor and is a measure of the damping behavior of the material. The modulus and tan δ depend on the temperature and the measuring frequency. At room temperature, rubbery materials show typical storage modulus values between 0.1 MPa and 10 MPa.

The curves in Figure 1.8 show the typical behavior of the storage and loss modulus of an amorphous and a semicrystalline polymer as a function of temperature.

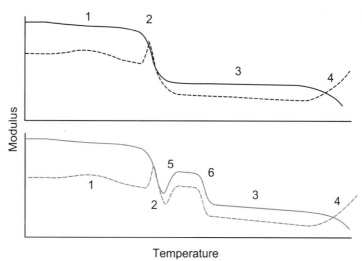

图 1.8　无定形聚合物（蓝色）和淬火冷却的部分结晶聚合物（红色）的储能模量 M′（实线）和损耗模量 M″（虚线）与温度的函数关系的典型曲线：1 次级松弛；2 主松弛；3 橡胶态（高弹态）；4 粘性流动；5 冷结晶；6 熔融。

Fig.1.8　Typical curves of the storage component (continuous line) and the loss component (dashed line) of the modulus as a function of the temperature for an amorphous polymer (blue) and a shock-cooled partially crystalline polymer (red):1 secondary relaxation; 2 main relaxation; 3 rubbery plateau; 4 viscous flow; 5 cold crystallization; 6 melting

在低温时，材料处于玻璃态，模量相对高（约 2GPa）。在该状态下常常

At low temperatures the material is in a glassy state. The modulus is relatively high (about 2GPa). In this state, a

测到次级转变(1),模量的实数部分在比较大的温度范围内稍稍降低,虚数部分显示一个宽平的峰。

在玻璃化转变区域(2)(也称为主松弛区域),储存模量在较窄温度范围内有几个数量级大小的变化,损耗模量显示尖锐的峰。样品从玻璃态转变为橡胶状的高弹态。

在该松弛区域,储能模量降低,损耗模量显示一个最大值。峰温依赖于频率,在较高测试频率下,松弛区域移向较高温度,移动的原因是由于松弛过程是分子重排的动力学过程。

主松弛区域后的一定温度区间内,模量几乎是恒定的。这个所谓的橡胶态(3)依赖于聚合度和交联。在更高温度下样品开始流动(4)(伴随着模量进一步降低)。

除了松弛行为,DMA 技术还测试半结晶材料中的相转变。结晶(5)和熔融(6)行为与松弛转变不同。一方面,模量实数部分的升降伴随着虚数部分相同的变化,此外,相应过程的温度范围不是依赖频率的。

除了所施加的应力(例如剪切或弯曲),在 DMA 测试中可变化的参数还有温度、频率和振幅。下表总结了从不同测试方法所能得到的信息。

secondary relaxation (1) is often measured in which the real part of the modulus decreases slightly over a relatively large temperature range and the imaginary part exhibits a broad flat peak.

In the region of the glass transition (2) (also known as the main relaxation region), the storage modulus then changes by several orders of magnitude over a relatively narrow temperature range. The loss modulus exhibits a distinct peak. The sample changes from a glassy to a rubbery-elastic state.

In the relaxation region, the storage modulus decreases and the loss modulus exhibits a maximum. The temperature of the peak maximum is frequency-dependent. At higher measuring frequencies, the relaxation region shifts to higher temperature. The reason for the shift is that relaxation behavior is determined by the dynamics of molecular rearrangements.

In a certain temperature interval after the main relaxation region, the modulus is then almost constant. This so-called rubbery plateau (3) is dependent on the degree of polymerization or cross-linking. At higher temperatures the sample begins to flow (4) (with further decrease of the modulus).

Besides relaxation behavior, the DMA technique can also measure phase transitions in semicrystalline materials. The behavior on crystallization (5) and melting (6) is different to that of a relaxation transition. For one thing, an increase or decrease of the real part of the modulus is accompanied by analogous behavior of the imaginary part. Besides this, the temperature ranges of the corresponding processes are not frequency-dependent.

Apart from the applied stress (e. g. shear or bending), the parameters that can be varied in DMA measurements are the temperature, frequency and amplitude. The following table summarizes the information that can be obtained from the different measurements methods.

温度变量 Variation of the temperature	频率变量 Variation of the frequency	振幅变量 Variation of the amplitude
● 松弛转变温度 Temperature of relaxation transitions ● 玻璃化转变 Glass transition ● 结晶 Crystallization ● 熔融 Melting ● 相容性 Compatibility ● 阻尼性能 Damping Behavior	● 松弛行为 Relaxation behavior ● 玻璃化转变 Glass transition ● 分子的相互作用 Molecular interaction ● 阻尼性能 Damping behavior	● 非线性力学行为 Non-linear mechanical behavior ● 填料效应 Effect of fillers

1.5 与 TGA 的同步测量　Simultaneous Measurements with TGA

在 1.2 节中提到，由于蒸发或在反应中产生的质量损失可用 TGA 高灵敏地测量。然而，测量结果的解释常常需要附加的信息，这可以通过两个或两个以上技术组合成一个仪器系统来获得。常用技术为同步 DTA 或同步 DSC（总是包含在 TGA/DSC 系统中）和各种在线气体分析方法（例如质谱和红外光谱）。

As already mentioned in Section 1.2, the loss of mass due to vaporization or in reactions can be detected with great sensitivity using TGA. The interpretation of the measurement results, however, frequently requires additional information. This can be obtained by combining two or more suitable techniques into one instrument system. Techniques often used are simultaneous DTA or simultaneous DSC (this is always included in a TGA/DSC system) and various methods for on-line gas analysis (e.g. mass spectrometry and infrared spectroscopy).

1.5.1 同步 DSC 和差热分析(DTA、SDTA)　Simultaneous DSC and Differential Thermal Analysis

同步 DSC 的测量原理和功能与 1.1.1 节中提到的常规 DSC 是一样的。

在差热分析(DTA)中，样品和参比物在炉体中被加热，样品和参比物之间的温度差用热电耦测量。如果样品中发生热效应（例如相转变或化学反应），额外的能量吸收或释放会改变样品的升温速率，这导致样品和参比物两边之间的温差。例如，在放热反应中，样品和参比物的温差大于反应前后。热效应用出现的台阶和峰来表示，正如在 DSC 测试中得到的曲线一样。

在 SDTA 中，没用参比样。程序温度相当于参比温度，样品温度是测量的。

同步 DSC 技术能在 TGA 实验中同步得到 DSC 曲线，DTA 或 SDTA 技术能在 TGA、TMA 和 DMA 实验中同步测量差热信号。这种信号常常有助于解释，因为该技术可测出并不伴随着质量或尺寸变化的热效应。例如，在 TMA 测试中，同步 SDTA 能够区分放热和

The measuring principle and functions of simultaneous DSC are same as those of a conventional DSC mentioned in Section 1.1.1.

In differential thermal analysis (DTA), a sample and a reference material are heated in a furnace. The temperature difference between the sample and the reference material is measured using thermoco-uples. If a thermal event occurs in the sample (such as a phase transition or chemical reaction), the additional uptake or release of energy changes the heating rate of the sample. This results in a temperature difference between the sample and reference sides. For example, during an exothermic reaction, the temperature difference between the sample and reference is larger than before or after the reaction. Thermal effects are indicated by the presence of steps and peaks, just as in a DSC measurement curve.

In SDTA (single DTA), no reference sample is used. The reference temperature corresponds to the program temperature and the sample temperature is measured.

The simultaneous DSC technique enables the DSC curve to be simultaneously measured in TGA experiments, and the DTA or SDTA technique enables the differential thermal signals to be simultaneously measured in TGA, TMA and DMA experiments. This often aids interpretation because it detects thermal events that are not accompanied by a change in mass or dimensions. For example, in TMA measurements, simultaneous SDTA can distinguish between exothermic and endothermic transitions, and

吸热转变,测出化学反应。 detect chemical reactions.

1.5.2 逸出气体分析(EGA) Evolved Gas Analysis

为了可靠解释 TGA 曲线,常常要了解 TGA 中样品逸出的气体性质。这些信息可通过将 TGA 仪器通过加热输送管连接到气体分析仪来获得,这能几乎同步地分析气体性质。两种最常用的在线联用是 TGA-MS(一台 TGA 偶联到一台质谱仪 MS)和 TGA-FTIR(一台 TGA 偶联到一台傅立叶转换红外仪 FTIR)。

For a reliable interpretation of TGA curves, one would often like to know more about the nature of the gases evolved from the sample in the TGA. This information can be obtained by connecting the TGA instrument to a gas analyzer by means of a heated transfer line. This allows the gases to be analyzed almost simultaneously. The two most frequently used on-line combinations are TGA-MS (a TGA coupled to a mass spectrometer, MS), or TGA-FTIR (a TGA coupled to a Fourier transform infrared spectrometer, FTIR).

这些技术的使用细节和应用见"热分析应用手册"的《热重-逸出气体分析》分册。

Practical details and applications of these techniques are given in "TGA-EGA" of "Application Handbooks Thermal Analysis".

1.5.2.1 TGA-MS

质谱仪(MS)由离子源、分析器和检测器组成。来自 TGA 的气体混合物(挥发性化合物、分解产物和保护气氛)在离子源中离子化生成分子离子和大量的碎片离子。离子按照质荷比(m/z)在分析器中被分开,然后在检测器系统中被记录。测得的离子谱显示出分子离子和大量从不同分子的碎裂形成的碎片离子。图谱或碎裂方式可表征所测试的特定化合物。此外,有些元素具有非常特殊的同位素波谱,如氯。

A mass spectrometer (MS) consists of an ion source, an analyzer and a detector. The gas mixture (volatile compounds/decomposition products and purge gas) arriving from the TGA is ionized in the ion source with the formation of molecular ions and numerous fragment ions. The ions are separated according to their mass-to-charge ratio (m/z) in the analyzer and then recorded by the detector system. The resulting mass spectrum displays the molecular ions and a large number of fragment ions formed from the fragmentation of the different molecules. The spectrum or fragmentation pattern is characteristic of the particular compound measured. Furthermore, some elements have very characteristic isotope patterns, e.g. chlorine.

一般来说,逸出气体可通过测得的质谱的碎裂方式来鉴定和表征。测得的图谱还可与图谱数据库中的图谱集对比。

In general, evolved gases can be identified or characterized by the fragmentation pattern of the mass spectra measured. The spectra obtained can also be compared with collections of spectra in spectral databases.

在与 TGA 联用中,并没有必要即通常并不在整个 TGA 测试范围内反复测试(即扫描)全部质量范围。可单独选择特定物质特有的质荷比(m/z)的碎片离子,记录其强度与时间或温度的函数关系。连续测试有限量特定碎片的这个技术称为多离子检测(MID)或选择离子监测

In combination with a TGA it is not always necessary or usual to measure (i.e. scan) the entire mass range repeatedly at short intervals throughout the TGA measurement. Fragment ions of mass-to-charge (m/z) ratio characteristic for particular substances can be individually selected and their intensities recorded as a function of time or temperature. This technique of continuously measuring a limited number of specific fragments is known as multiple ion detection (MID) or selected ion monitoring

(SIM),可促使灵敏度大大提高。
MS 离子曲线与 TGA（或 DTG）曲线的比较能鉴定或确认热效应中失重物质的存在。

1.5.2.2 TGA—FTIR

红外光谱法也是用来鉴定 TGA 分析中形成的气体产物。TGA 仪器通过加热输送管与傅立叶转换红外光谱仪（FTIR）直接偶联，测量自 TGA 仪器到达 FTIR 气体池的气体混合物（挥发性化合物、分解产物和吹扫气体）在 4000 至 400cm^{-1}（波数）范围内的红外图谱。与分子中不同键的特定振动频率对应的波长处的红外能量被吸收，这能鉴定官能团和提供关于所得相关物质的性质信息。

现代 FTIR 仪器能在几秒内测试高质量的图谱，在整个 TGA 分析范围内连续记录图谱，这意味着实际上对 TGA 分析的任何温度或时间点都可获得单独谱图，这些谱图可与数据库参比谱图作比较，进行评估。

在实际中，为了简化数据处理，对利用了谱图中包含了全部或仅是部分信息的随时间（或温度）变化的曲线进行记录，这些曲线称为 Gram-Schmidt 曲线或化学谱。

简单说来，Gram-Schmidt 曲线是通过对每个单独图谱的吸收波带强度积分得到的，该信息表示为吸收强度与时间函数关系的曲线。曲线是任何瞬间到达 FTIR 仪器气体池内的物质量的量度。该曲线一般与 DTG 曲线相似，除了由于气体在连接两个仪器的传输管内输送产生的 DTG 和 Gram-Schmidt 峰之间一个明显的短滞后（通常只有几秒）。因为单独物质的红外吸收强度很不同，所以强度没有直接可比性。

(SIM) and results in a large increase in sensitivity.
A comparison of the MS ion curves with the TGA (or DTG) curve allows the presence of a substance responsible for the mass loss in a thermal effect to be identified or confirmed.

Infrared spectroscopy can also be used to identify gaseous products formed in a TGA analysis. The TGA instrument is coupled directly to a Fourier transformation infrared spectrometer (FTIR) by a heated transfer line. Infrared spectra of the gas mixture (volatile compounds / decomposition products and purge gas) arriving in the FTIR gas cell from the TGA instrument are measured in the range 4000 to 400cm^{-1} (wavenumbers). Infrared energy is absorbed at wavelengths that correspond to the specific vibrational frequencies of the different bonds in the molecules. This allows functional groups to be identified and information on the nature of the substances involved to be gained.

A modern FTIR spectrometer is capable of measuring good quality spectra within a few seconds. The spectra are recorded continuously throughout the TGA analysis. This means that individual spectra are available for practically any temperature or point in time of the TGA analysis. These spectra can be compared with database reference spectra and evaluated.

In practice, to simplify the evaluation, curves are recorded as a function of time (or temperature) that make use of all or just part of the information included in the spectrum. These curves are known as Gram-Schmidt curves and chemigrams.

In simple terms, the Gram-Schmidt curve is obtained by integrating the absorption band intensities of each individual spectrum. The information is presented as a curve of absorption intensity plotted as a function of time. The curve is a measure of the quantity of substance arriving in the gas cell of the FTIR spectrometer at any instant. The curves often resemble DTG curves except that there is obviously a short delay (normally just a few seconds) between the DTG and the Gram-Schmidt peaks due to the gas transport in the transfer line connecting the two instruments. The intensities are not directly comparable because individual substances have very different infrared absorption intensities.

化学谱显示了特定波数区域的红外吸收与时间(或温度)的函数关系,能观察气体产物的形成(通过官能团)与时间的函数关系。例如,3640~3620 cm^{-1}(O—H 振动)和 1250~1000cm^{-1}(C—O 振动)范围的化学谱是乙醇特有的。用示意图例举于图 1.9 中。在第一个失重台阶,是乙醇逸出的过程,如同所预期的,两个化学谱都显示一个峰。在第二个失重台阶,只有化学谱 B 显示一个峰。在这里,一个不同类的物质逸出,要鉴定它需要另外的化学谱。随后,通过 DTG 峰期间 FTIR 图谱的详细检查,能得到更准确的分析。

Chemigrams display the infrared absorption of particular wavenumber regions as a function of time (or temperature and allow the formation of gaseous products (via functional groups) to be observed as a function of time. For example, chemigrams of the regions 3640-3620 cm^{-1}(O—H vibration) and 1250—1000 cm^{-1}(C—O vibration) are specific for alcohols. An example is shown schematically in Figure 1.9. In the first weight loss step, a process in which an alcohol is evolved, both chemigrams show a peak, as expected. In the second weight loss step, only chemigram B shows a peak. A different type of substance is evolved here and additional chemigrams are required to identify it. A more accurate analysis can afterward be obtained through a detailed examination of the FTIR spectra measured during the DTG peaks.

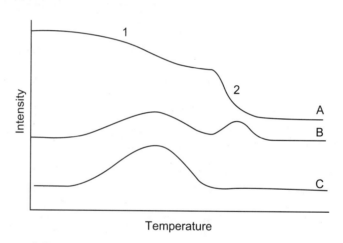

图 1.9 显示 TGA 曲线(A)和两个化学谱(B:3640—3620cm^{-1};C:1250—1000cm^{-1})的示意图。

Fig. 1.9 Schematic diagram showing a TGA curve (A) and two chemigrams (B: 3640—3620 cm^{-1}; C: 1250—1000 cm^{-1}).

2 聚合物的结构和性能
Structure and behavior of polymers

2.1 聚合物领域的一些定义
Some Definitions in the Field of Polymers

聚合物是很大的有机分子,通常称为大分子。它们的分子量通常大于 10000 g/mol,大至几乎无限高的值。许多天然产物如纤维素、淀粉和蛋白质是聚合物。从化学上说,在过去 100 年里一直在制造聚合物。在聚合反应中,单体以一定的序列连接起来,就合成了大分子。单体是小的具有反应性的分子,在常压条件下通常是气体或液体。每个大分子的单体单元数与聚合程度有关。单一的单体聚合成均聚物,例如,乙烯形成聚乙烯均聚物。

Polymers are very large organic molecules, so-called macromolecules. Their molar mass is usually greater than 10000 g/mol and ranges upward to almost infinitely high values. Many natural products such as cellulose, starch and proteins are polymers. Polymers have been made "chemically" for the last 100 years. These macromolecules are synthesized when monomers join together in certain sequences in a polymerization reaction. Monomers are small reactive molecules, usually gaseous or liquid under ambient conditions. The number of monomer units per macromolecule is related to the degree of polymerization. A pure monomer polymerizes to a hom-opolymer, e.g. ethylene forms a polyethylene homopolymer.

$$(CH_2-CH_2-)_n$$

混合的单体例如乙烯和丙烯还形成共聚物,此例为乙烯－丙烯共聚物,不同单体单元随机分布生成所谓的无规共聚物。嵌段共聚物由一定数量的同一单体链段与其它单体链段相连组成。最后,接枝共聚物由支化侧链接枝在大分子上生成。

Mixed monomers, e.g. ethylene and propylene, also form copolymers, in this case an ethylene-propylene copolymer. A random distribution of the different monomer units leads to so-called random copolymers. The block copolymer consists of a certain number of identical monomer blocks followed by blocks of the other monomer. Finally, a graft copolymer is made by grafting side chains on a macromolecule.

与金属相比,聚合物的性能高度依赖于温度。因此,热分析对表征和鉴定聚合物是一组十分有用的技术。如同一切有机物质,聚合物在高温下分解——特别当有氧存在时。

Compared with metals, the properties of polymers are strongly dependent on temperature. For this reason, thermal analysis is a very useful group of techniques for the characterization and identification of polymers. As with all organic matter, polymers decompose at elevated temperature — especially in presence of oxygen.

聚合物分为三类:
- 热塑性塑料,像聚乙烯和聚苯乙烯这样的塑料
- 弹性体,像天然橡胶和丁苯橡胶这样的橡胶
- 热固性树脂,像不饱和树脂和环氧树脂

Polymers are classified into three groups:
- Thermoplastics, "plastics" such as polyethylene and polystyrene
- Elastomers, "rubber" such as natural rubber and styrene-butadiene rubber
- Thermosets, such as unsaturated polyesters and epoxy resins.

Polymer mixtures (so-called blends) are also frequently used because this allows many important polymer properties to be optimized. Physical properties are also influenced by the orientation of the macromolecules as a result of molding, e. g. injection molding or fiber spinning. Last but not least, quality is influenced through the use of a wide range of additives.

Examples are:
- Fillers such as chalk or carbon black
- Reinforcing materials such as glass or carbon fibers
- Plasticizers
- Blowing agents
- Antioxidants, UV and heat stabilizers
- Flame retardants
- Dyestuffs, pigments
- Mold lubricants

2.2 Physical Structure of Polymers

At higher temperatures polymers exist as amorphous melts. This means that the macromolecules are present as trapped entanglements. The individual segments of the polymer chain exhibit a relatively large degree of mobility. If the material is cooled so rapidly that it cannot crystallize, a strongly supercooled amorphous melt is formed that on further cooling passes through the glass transition. An amorphous glass is formed. On cooling, molecular mobility decreases and the viscosity of the material increases. The supercooled melt has often a highly viscous and frequently sticky consistency. At the glass transition, the viscosity increases by several decades. In the glassy state the material has the structure of an amorphous liquid but the mechanical properties of a solid. The reason for this behavior is that, at the glass transition, the cooperative molecular rearrangements typical for a liquid "freeze" so that only the vibrations typical for a solid remain. In practical terms, this means that below the glass transition temperature the polymer is hard and relatively brittle.

Polymers stored between their melting temperature and glass transition temperature can undergo crystallization. Due to their molecular size and the chemically incorporated defects (chain ends and chain branching), the polymer crystallites so formed differ markedly from the crystals of substances of low molecular

不同。微晶是微小的、未完全形成的晶体。在结晶过程中,伸长的分子被强制折叠回自身(链折叠)。折叠链常常在无定形物质或不能再结晶的物质之间以一层层薄层存在。晶体层和无定形层交替,形成薄层堆积。聚合物以部分结晶和部分无定形状态存在的,被称为半结晶聚合物。所述的聚合物结构示意于下图。

weight. Crystallites are small, imperfectly formed crystals. In the crystallization process, the elongated molecules are forced to fold back on themselves (chain folding). The folded chains are often present as layers of lamellas between which amorphous material, or material that is no longer able to crystallize, is present. Crystalline layers and amorphous layers alternate and form lamellar stacks. The polymers are partly crystalline and partly amorphous and are referred to as semicrystalline polymers. The polymer structures mentioned are shown schematically in the following diagram.

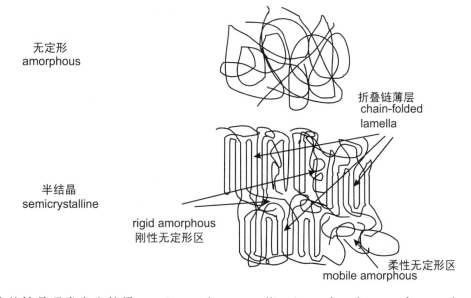

由于聚合物的结晶通常发生较慢,所以能通过快速冷却抑制至玻璃化转变温度以下。根据经验规则,玻璃化转变温度大约等于熔融温度的三分之二,用绝对温标测量。这个规则适用于较大的微晶。如果一个能够结晶的无定形聚合物被加热到玻璃化转变温度以上,它能在相对低的温度下结晶。这类结晶也称为冷结晶。

聚合物微晶的熔融温度取决于晶体尺寸。厚层比薄层在更高温度下熔融。层的厚度不仅取决于结晶温度,而且取决于链的化学结构。在这个方面,支链和其长度之间的距离是非常重要的因素。对于侧链长度和间距分布宽的聚合物,会出现聚合物微晶尺寸的宽范围分布。在这种情况下,在玻璃化转变温度以

Since the crystallization of polymers frequently takes place relatively slowly, it can be suppressed by cooling rapidly to below the glass transition temperature. A rule of thumb says that the glass transition temperature is about two-thirds of the melting temperature-measured in Kelvin. This rule applies to relatively large crystallites. If an amorphous polymer that is capable of crystallizing is heated to temperatures above the glass transition temperature, it can crystallize at relatively low temperatures. This type of crystallization is also known as cold crystallization.

The melting temperature of the polymer crystallites depends on the crystal size. Thick lamellas melt at higher temperatures than thin ones. The thickness of the lamella depends not only on the crystallization temperature but also on the chemical structure of the chain. In this respect the distance between chain branches and their length is a very important factor. With polymers having a wide distribution of side chain lengths and distances, a wide range of distribution of the size of the polymer crystallites can occur. In this case, very small and unstable crystallites that melt shortly above the

上很快就熔融的很小的和不稳定的微晶会形成较大的稳定的微晶。在这些情况下,可观察到宽的熔融范围,这类行为能在聚乙烯(PE)中看到。

除了微晶本身外,半结晶聚合物也含有无定形区。由于活动程度不同,还分成柔性无定形区和刚性无定形区。刚性无定形区主要在微晶的折叠表面。在这些区域活动是受限的,因此它们不参与玻璃化转变。

也有聚合物由于其化学结构的原因而根本不能结晶的。这些聚合物永远是无定形的。一个很好的例子是聚苯乙烯(PS),苯乙烯侧链基团阻止了结晶。

glass transition temperature can form next to relatively large stable crystallites. In such cases, a wide melting range is observed. This type of behavior can be observed with polyethylene (PE).

Apart from the crystallites themselves, semicrystalline polymers also contain amorphous regions. Because of different degrees of mobility, one distinguishes between mobile amorphous and rigid amorphous regions. The rigid amorphous regions are found primarily at the fold surfaces of the crystallites. Movement is so restricted in these regions that they do not participate in the glass transition.

There are also polymers that are unable to crystallize at all because of their chemical structure. These polymers are always amorphous. A good example is polystyrene (PS), where the styrene side chain groups prevent crystallization.

2.3 热塑性聚合物 Thermoplastic Polymers

热塑性聚合物在加热时熔融或流动。它们由无规缠结的(无定形热塑性塑料)或以微晶方式部分有序的(半结晶热塑性塑料)线性大分子组成。

注:本书中提到的聚合物的所有名称和符号都依照 ISO 1043-1。

Thermoplastic polymers melt or flow on heating. They consist of linear macromolecules that are either randomly coiled (amorphous thermoplastics) or partly ordered as crystallites (semicrystalline thermoplastics).

Note: All the names and symbols of polymers mentioned in this booklet are written according to ISO 1043-1.

2.3.1 无定形塑料 Amorphous Plastics

在软化点温度(玻璃化转变温度即玻璃化温度)以下,无定形聚合物是坚硬的或刚性的;有些是脆性的,例如聚苯乙烯。加热时,在玻璃化转变温度软化,成为粘弹性的。进一步加热时,粘度降低,并无确定熔点。玻璃化转变温度受单体的化学结构和聚合度的影响,因此用玻璃化转变温度来表征或鉴定聚合物。然而,增塑剂的存在能降低玻璃化转变温度,水分也可充当增塑剂的作用。在玻璃化转变处,聚合物的比热增加约 0.1 至 0.4 J/gK,生成

Below their softening temperature (glass transition temperature or glass temperature), amorphous polymers are stiff or rigid; some are brittle, e. g. polystyrene. On heating, they soften at their glass transition temperature and become viscoelastic. On further heating, the viscosity decreases without a distinct melting point becoming apparent. The glass transition is influenced by the chemical structure of the monomer and the degree of polymerization. Polymers can therefore be characterized or identified by their glass transition temperature. The glass transition temperature is however lowered by the presence of plasticizers. Moisture can also act as a plasticizer. At the glass transition, the specific heat capacity of a polymer increases by some 0.1 to 0.4 J/gK causing a step in the DSC

DSC 曲线上的一个台阶。通常,在第一轮加热时可看到一个吸热峰(称为松弛峰)。这是由在低于玻璃化转变温度时,贮存期间发生的焓松弛所引起的。

玻璃化转变温度的测量是热分析中最经常的应用之一。主要物理性能与温度的函数关系的变化列于下图中。

curve. Often, an endothermic peak (a so-called relaxation peak) is visible in the first heating run. This is caused by enthalpy relaxation occurring during storage at temperatures below the glass transition temperature.

Measurement of the glass transition temperature is one of the most frequent applications in thermal analysis. The changes of the main physical properties as a function of temperature are shown in the following diagram.

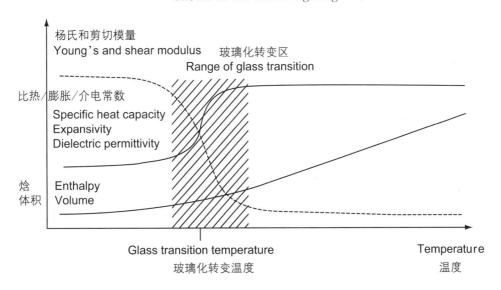

2.3.2 半结晶塑料 Semicrystalline Plastics

半结晶塑料在到达结晶开始熔化的温度前是坚硬的,例如聚对苯二甲酸乙二醇酯。只有规整的大分子能够结晶,无定形区域也存在于微晶之间。因此,不存在像金属或盐那样的有 100% 结晶度的物质。结晶度(晶态物质的百分数)取决于聚合物分子的规整度和结晶发生时的条件。添加成核剂和结晶抑制剂影响最终的结晶度。硬度和强度随结晶度的增大而增大。

结晶度通过将特定样品的熔融热(由 DSC 测定)除以该 100% 结晶物质的理论熔融热得到。结晶度值通常在 20% 至 75% 范围内。

塑料的"熔融行为"(熔程、熔融分数与温度的函数关系、熔融温度)对其加工性和鉴定十分重要。熔融温度

Semicrystalline plastics remain rigid until temperatures are reached at which the crystallites begin to melt, e.g. polyethylene terephthalate. Only regular macromolecules are able to crystallize. Amorphous regions are also present between the crystallites. For this reason there is no such thing as 100% crystallinity as is the case with metals or salts. The degree of crystallinity (percentage of the material in the crystalline state) depends on the regularity of the polymer molecules and on the conditions under which crystallization occurs. The additions of nucleating agents or crystallization inhibitors influence the final degree of crystallinity. Hardness and strength increase with increasing crystallinity.

The degree of crystallinity is obtained by dividing the heat of fusion (determined by DSC) of a particular sample by the theoretical heat of fusion of the 100% crystalline material. The value usually ranges from 20 to 75%.

The "melting behavior" (melting range, fraction melted as a function of temperature, melting temperature) of a plastic is of great importance for its processability as well as its identification.

是最后的微晶熔化的温度,大约等于 DSC 的峰温。熔程取决于微晶的尺寸分布——最小的微晶最先熔化。DSC 曲线上几个峰的存在也可能表示多晶态,即存在不同晶体变型,例如聚酰胺。另一个经常观察到的效应是熔融中断,这由先前的退火所致。

微晶在熔融时被破坏,但通常在冷却时重新形成。熔融聚合物在冷却时的再结晶行为显示其结晶的倾向(过冷和晶体生长速率)。因此,许多通常为半结晶的聚合物能通过快速冷却熔融塑料至其玻璃化温度以下将其冻结在无定形状态。这被称为骤冷或淬火冷却。

半结晶塑料的共混物,还有嵌段共聚物,生成与用单独熔融曲线叠加形成的曲线相似的 DSC 熔融曲线,这能鉴定单一的塑料。

The melting temperature is the temperature at which the very last crystallites melt and is approximately equal to the DSC peak temperature. The melting range is determined by the size distribution of the crystallites-the smallest crystallites melt first. The presence of several peaks in the DSC curve can also indicate polymorphism, i. e. the presence of different crystal modifications, e. g. in polyamides. Another effect frequently observed is the melting gap, which is caused by previous annealing.

The crystallites are destroyed on melting but usually reform on cooling. The recrystallization behavior of molten polymers on cooling indicates their tendency to crystallize (supercooling and crystal growth rate). Many normally semicrystalline polymers can therefore be frozen in the amorphous state by rapidly cooling the molten plastic to below its glass temperature. This is referred to as shock-cooling or quench-cooling.

Blends of semicrystalline plastics, and also block copolymers, yield DSC melting curves that resemble curves formed from the superposition of individual melting curves. This allows the individual plastics to be identified.

3 热塑性聚合物的重要领域
Important Fields of Thermoplastic Polymers

行业 Branch	聚合物的特有优势 Special advantages of the polymers	代表性产品 Typical products	应用聚合物 Applied polymers
农业 Agriculture	耐候、防腐、易成型 weather-resistant, corrosion-proof, ease of molding	覆盖薄膜、温室、贮料垛、软管、管、用具、绝热材料 cover films, greenhouses, silos, hoses, pipes, equipment, heat insulation material	PE、PP、PVC、PMMA
汽车工业 Automotive industry	低密度、防腐蚀、由于振动阻尼而静音、易成型 low density, corrosion resistant, silent due to vibration damping, ease of molding	点火专门部件、电池、灯具、柄、方向盘、安全带和轮胎中的纤维、燃料箱、粘合剂、轴承、漆、管道系统 technical parts for ignition, battery, lights, handles, steering wheel, fibers in safety belts and tires, fuel tanks, adhesives, bearings, coatings, piping	PA、POM、PMMA、PVC、PE、PET、ABS
航空业 Aviation	低密度、由于振动阻尼而静音 low density, silent due to vibration damping	衬料、柄、粘合剂、灯、窗玻璃、密封材料 linings, handles, adhesives, lights, window glazing, sealing material	PA、PE、ABS、PMMA
建筑工业 Building industry	防腐蚀、易成型、美观怡人 corrosion resistant, easily molded, aesthetically pleasing	薄膜、室内地面材料、窗、卫生设备、管道、热绝缘材料、粘合剂、油漆、穹顶灯 films, flooring materials, windows, sanitary installations, piping, heat insulation, adhesives, paints, domed lights	PVC、PE、PS、PVAC、PMMA
电气工业 Electrical industry	高电阻、低损耗因子 high electrical resistance, low loss factor,	电缆绝缘材料、管、设备部件、箱、键盘、衬料、电容器薄膜 cable insulation, pipes, parts of equipment, cases, keyboards, linings, films for capacitors	PE、ABS、PVC、PS
家用 Household	表面光滑、防腐、无嗅无味 smooth surface, corrosion-proof, odorless, tasteless	厨房用具、烹饪袋、浴室用具、家具、玩具 kitchen appliances, cooking bags, bathroom utensils, furniture, toys	PE、PP、PA、PVC、ABS、PS、PMMA
休闲和体育用品 Leisure and sporting goods	高强度、抗磨损、耐气候、易着色 high tensile strength, wear resistant, weather resistant, easily colored	船、帆、溜冰和滑雪靴、滑翔伞、球、轻便折叠躺椅 boats, sails, skis and skiing boots, paragliders, balls, deckchairs	PET、PE、PP、ABS、PA、PVC
机器和仪器制造 Machine and instrument construction	抗磨损、防腐、因振动阻尼而静音、易成型 wear resistant, corrosion resistant, silent due to vibration damping, easily molded	轴承、驾驶盘、罩、箱、板、衬料、漆、防腐层、管道、泵和风扇 bearings, gear wheels, housings, cases, sheets, linings, coatings, corrosion protecting, tubing, pumps and fans	PA、POM、PC、PVC、PTFE

(续表)

行业 Branch	聚合物的特有优势 Special advantages of the polymers	代表性产品 Typical products	应用聚合物 Applied polymers
医药 Medicine	表面光滑、能消毒、惰性 smooth surface, can be sterilized, inert	管、注射器、袋、设备 pipes, syringes, bags, equipment	PE、PP、PVC、PS、PC
包装 Packaging	易成型、表面光滑、重量轻、易着色、可印刷 easily molded, smooth surface, low weight, easily colored, printability	薄膜(及食品复合膜)、袋、瓶、容器 films (also composite films for food), bags, bottles, containers	PE、PP、PVC、PET、PS
纺织工业 Textile industry	高强度、抗磨损、耐气候、易着色 high tensile strength, wear resistant, weather resistant, easily colored	纤维、线、绳、无纺布、袜、鞋 fibers, threads, ropes, non-woven fabrics, stockings, shoes	PET、PA、PAN、PE、PP、PVC、CA

4 热塑性聚合物的应用一览表
Application Overview of Thermoplastic Polymers

该表显示了可用热分析进行研究的性质，重要的应用用较大的圈标记。

The table shows the properties that can be investigated by TA. Important applications are marked with a large circle.

	DSC	TMA	TGA	DMA
熔融温度 Temperature of fusion	●	●		·
熔融热 Heat of fusion	●			
结晶度 Crystallinity	●			
熔融行为,熔融分数 Melting behavior, fraction melted	●	·		·
结晶温度 Temperature of crystallization	●			
结晶热 Heat of crystallization	●			
冷结晶 Cold crystallization	●	·		·
多晶体(晶体变型) Polymorphism (change of crystal modification)	●	●		·
玻璃化转变 Glass transition	●	●		●
软化 Softening		●		·
挥发、解吸附（水分）、蒸发 Evaporation, desorption (moisture), vaporization	●		●	
热分解（热解、解聚）Thermal decomposition (pyrolysis, depolymerization)	·	·	●	
热稳定性 Thermal stability	·	·	●	
氧化降解、氧化稳定性 Oxidative degradation, oxidation stability	●	·	·	
成分分析(挥发物、聚合物、碳黑、灰分、填料、玻璃纤维) Compositional analysis (volatiles, polymer, carbon black, ash, filler, glass fibers)			●	
比热 Specific heat capacity	●			
粘弹行为 Viscoelastic behavior		·		●
膨胀和收缩行为 Expansion and shrink behavior		●		
膨胀性(线性膨胀系数) Expansivity (coefficient of linear expansion)		●		
杨氏和剪切模量,刚度 Young's and shear modulus, stiffness		·		●
阻尼性能 Damping behavior				●

5 热塑性聚合物的特征温度表 Table of Characteristic Temperatures of Thermoplastic polymers

此表可用于鉴别热塑性聚合物。它汇总了典型的玻璃化转变温度 T_g、DSC 峰温 T_m、微晶的熔融热 $\triangle H_{fus}$。

聚合物根据温度增高排列。括号中的值不确定。

This table can be used to identify thermoplastic polymers. It summarizes typical glass transition temperatures, Tg, DSC peak temperatures, Tm, and crystallite heat of fusion, $\triangle H_{fus}$.

The polymers are arranged according to increasing temperatures. Values in brackets are uncertain.

聚合物 Polymer	符号 Symbol	T_g ℃	T_m ℃	$\triangle H_{fus}$,100%结晶 100% cryst. J/g
聚氯乙烯,增塑 Polyvinyl chloride, plasticized	PVC-P	—40⋯10		
乙烯/醋酸乙烯共聚物 Ethylene-vinyl acetate copolymer	E/VAC	—20⋯20	40⋯100	
聚乙烯,低密度 Polyethylene, low density	PE-LD	(—100)	110	293
聚乙烯,高密度 Polyethylene, high density	PE-HD	(—70)	135	293
聚甲醛共聚物 Polyoxymethylene copolymer	POM		164⋯168	
聚丙烯 Polypropylene	PP	(—30)	165	207
聚偏二氯乙烯 Polyvinylidene chloride	PVDC	—17		
聚甲醛均聚物 Polyoxymethylene homopolymer	POM		175⋯180	326
聚偏二氟乙烯 Polyvinylidene fluoride	PVDF		178	
聚酰胺 12 Polyamide 12	PA12	(40)	180	
聚酰胺 11 Polyamide 11	PA11		186	
聚醋酸乙烯 Polyvinyl acetate	PVAC	30		
聚氯乙烯 Polyvinyl chloride	PVC-U	80⋯85	(190)	
聚对苯二甲酸丁二醇酯 Polybutylene terephthalate	PBT	65	220	
聚酰胺 6 Polyamide 6	PA6	(40)	220⋯230	230
聚酰胺 610 Polyamide 610	PA610	(46)	226	
聚乙烯醇 Polyvinyl alcohol	PVAL	85		

(续表)

聚合物 Polymer	符号 Symbol	T_g ℃	T_m ℃	ΔH_{fus},100％结晶 100％cryst. J/g
聚苯乙烯 Polystyrene	PS	90…100		
聚甲基丙烯酸甲酯 Polymethyl methacrylate	PMMA	105		
聚苯醚 Polyphenylene oxide	PPE		230	
聚碳酸酯 Polycarbonate	PC	155	(235)	
聚酰胺66 Polyamide 66	PA66	(50)	260	255
聚对苯二甲酸乙二醇酯 Polyethylene terephthalate	PET	69	256	140
乙烯/四氟乙烯共聚物 Ethylene/tetrafluorethylene copolymer	E/TFE		270	
氟乙烯/丙烯共聚物 Fluorethylene/propylene copolymer	FEP		280	
聚苯硫醚 Polyphenylene sulfide	PPS	80	280	
聚邻苯二甲酰胺 Polyphthalamide	PPA	125	310	
聚丙烯腈 Polyacrylonitrile	PAN	100	(320)	
聚醚醚酮 Polyetheretherketone	PEEK	143	335	
聚四氟乙烯 Polytetrafluorethylene	PTFE	(−20)	327	82
聚醚砜 Polyether sulfone	PES	220		
聚醚亚胺 Polyetherimide	PEI	220		

(续表)

6 重要热塑性聚合物的性能和典型的热分析应用
Properties of Important Thermoplastic Polymers and Typical TA Applications

6.1 聚乙烯,PE Polyethylene

聚乙烯 PE 是半结晶热塑性聚合物。不同生产方法生产的聚合物的分子结构和密度不同。高密度聚乙烯 PE-HD 实际上支链很少,因此其结晶度高(60 至 80%),密度约 0.95g/cm³。低密度聚乙烯 PE-LD 是高度支化的,具有较低的结晶度(20 至 40%),密度约 0.92 g/cm³。在 PE-LLD(线性低密度聚乙烯)中,结晶被与 α-烯烃共聚合所产生的短侧链抑制,从而获得透明性好(对包装用途重要)的高强度薄膜。
PE 是低吸水的非极性材料。它是出色的电绝缘体。成型后的聚乙烯能通过交联增强尺寸稳定性。
PE 的热分析:
DSC: 通过熔融行为、结晶度、氧化稳定性来表征。
TGA: 降解行为、碳黑含量、填料含量。
TMA: 膨胀行为、膨胀性、交联度。
DMA: 模量、粘弹和阻尼性能、玻璃化转变。

Polyethylene, PE, is a semicrystalline thermoplastic polymer. Different production methods are used to produce particular qualities of macromolecular structure and density. The macromolecules of high-density polyethylene, PE-HD, are practically unbranched. The degree of crystallinity is therefore high (60 to 80%) and the density is about 0.95 gcm^{-3}. Low-density polyethylene, PE-LD, is strongly branched and has a lower crystallinity (20 to 40%) and a density of about 0.92 gcm^{-3}. In PE-LLD (linear low density) the crystallization is hindered by short side groups created by copolymerization with α-olefins, e.g. hexene to obtain a film of high strength combined with good transparency (important for packaging purposes).
PE is a nonpolar material with low water sorption. It is an excellent electrical insulator. Polyethylene after molding can be cross-linked to enhance dimensional stability.
Thermal analysis of PE:
DSC: Characterization via melting behavior, crystallinity, oxidation stability.
TGA: Degradation behavior, carbon black content, filler content.
TMA: Expansion behavior, expansivity, degree of crosslinking.
DMA: Modulus, viscoelastic and damping behavior, glass transition.

6.2 乙烯/醋酸乙烯共聚物, E/VAC Ethylene/Vinylacetate Copolymer

这两个单体能以任何比例共聚。随着 VAC 含量增大,材料更软和更透明。性能在半结晶热塑性到无定形、橡胶态、良好低温性能的范围内变化。E/VAC 是最早的商品化热塑性弹性体(易成型,性能几乎类似硫化橡胶)。
应用:软管和软水管、薄膜(深冻袋)、耐热和耐候"橡胶"制品、热熔胶。

These two monomers can be copolymerized in any ratio. With increasing VAC content, the material becomes softer and more transparent. The properties range from being semicrystalline thermoplastic to amorphous, rubber-like and good low temperature properties. E/VAC was the first commercially available thermoplastic elastomer (easily molded, properties almost like a vulcanized rubber).

Applications: soft tubes and hoses, films (bags for deep freezer), heat and weather resistant "rubber" ware, hot melt.

E/VAC 的热分析：
DSC：通过玻璃化转变（约 −40℃）和熔程（40 至 100℃）行为、氧化稳定性来表征。
TGA：热稳定性、降解行为、填料含量。
TMA：玻璃化转变、膨胀行为、膨胀性、热塑性流动。
DMA：模量、粘弹和阻尼性能、玻璃化转变。

Thermal analysis of E/VAC：
DSC： Characterization via glass transition (approx. −40℃) and melting range (from 40 to 100℃), oxidation stability.
TGA： Thermal stability, degradation behavior, filler content.
TMA： Glass transition, expansion behavior, expansivity, thermoplastic flowing.
DMA： Modulus, viscoelastic and damping behavior, glass transition.

6.3　聚丙烯，PP　Polypropylene

常规合成的 PP 大分子生成半结晶材料。PP 比 PE 更硬，尺寸稳定性更高，甲基的存在使其更易氧化。
应用：薄膜、电缆绝缘材料、热水管、家用电器、实验室设备和洗衣机部件、绳。
PP 的热分析：
DSC：通过熔融行为、结晶度、氧化稳定性来表征。
TGA：降解行为、碳黑含量、填料含量。
TMA：膨胀行为、膨胀性。
DMA：模量、粘弹和阻尼性能、玻璃化转变。

Regularly synthesized PP macromolecules result in a semicrystalline material. PP is harder and has a higher dimensional stability compared with PE. But it is prone to oxidation due to its methyl groups.
Applications：films, cable insulation, hot water pipes, household appliances, parts of laboratory equipment and washing machines, ropes.
Thermal analysis of PP：
DSC： Characterization via melting behavior, crystallinity, oxidation stability.
TGA： Degradation behavior, carbon black content, filler content.
TMA：Expansion behavior, expansivity.
DMA： Modulus, viscoelastic and damping behavior, glass transition.

6.4　聚苯乙烯，PS　Polystyrene

聚苯乙烯 PS 是无色、透明、坚硬和脆性的。为了改善抗冲击性，生产了二元共聚物和三元共聚物【苯乙烯-丙烯腈共聚物：SAN（US 和日本：AS）；苯乙烯-丁二烯共聚物：S/B；丙烯腈-丁二烯-苯乙烯三元共聚物：ABS】。SAN 共聚物是透明的，呈浅黄色；S/B 的表面较暗；ABS 不透明。
PS 的应用：廉价的透明大宗产品、客户礼品、一次性杯、玩具、包装和绝缘发泡 PS。

Polystyrene, PS, is colorless, transparent, rigid and brittle. For better impact resistance copolymers and terpolymers are produced [styrene-acrylonitrile copolymer：SAN (US and J：AS); styrene-butadiene copolymer：S/B, acrylonitrile-butadiene-styrene terpolymer：ABS]. SAN copolymers are transparent with a light yellowish tone, the surface of S/B is dull, ABS is opaque.

Applications of PS：cheap transparent mass products, customer gifts, one way beakers, toys, foamed PS for packaging and heat insulation.

高抗冲共聚物和三元共聚物:冰箱的内胆、真空吸尘器的外罩、玩具、安全帽、小船。

聚苯乙烯塑料的热分析:

DSC: 通过玻璃化转变(约100℃,与丁二烯的共聚物约−85℃,与丙烯腈的共聚物(约135℃)和热稳定性(解聚开始)来表征。

TGA: 热稳定性、降解行为、填料含量。

TMA: 玻璃化转变、膨胀行为、膨胀性、热塑性流动。

DMA: 模量、粘弹和阻尼性能、玻璃化转变。

High impact copolymers and terpolymers: inner linings of refrigerators, housings of vacuum cleaners, toys, safety helmets, small boats.

Thermal analysis of polystyrene plastics:

DSC: Characterization via glass transition (approx. 100℃, copolymer with butadiene in addition approx. −85℃, with acrylonitrile in addition (approx. 135℃), thermal stability (onset of depolymerization).

TGA: Thermal stability, degradation behavior, filler content.

TMA: Glass transitions, expansion behavior, expansivity, thermoplastic flowing.

DMA: Modulus, viscoelastic and damping behavior, glass transition.

6.5 聚氯乙烯,PVC　Polyvinyl Chloride

PVC是无定形的、透明的(但常常由于填料而不透明)、化学稳定的。进一步氯化(PVCC)将玻璃化温度从大约85℃提高到100℃。另一方面,通过添加增塑剂(PVC-S)能降低到几乎任意值。最大的弱点是始于约180℃的热降解,所释放的盐酸具有强的腐蚀性。

应用:

PVC-U: 窗框、废水管、汽车内层、仪表盘、酸容器、护墙板。

PVC-S: 管材、玩具、球、屋顶防水和游泳池用薄膜、人造革。

PVC的热分析:

DSC: 通过玻璃化转变(约−40℃~+90℃,取决于增塑剂含量)和热稳定性来表征。

TGA: 热稳定性、降解行为、填料含量。

TMA: 玻璃化转变、膨胀行为、膨胀性。

DMA: 模量、粘弹和阻尼性能、玻璃化转变、凝胶化。

PVC is amorphous, transparent (but frequently opaque due to fillers) and chemically stable. Further chlorination (PVCC) increases the glass temperature from approx. 80℃ to 100℃. On the other hand, it can be lowered to almost any value by the addition of plasticizers (PVC-S). The great disadvantage is thermal degradation, which begins around 180℃. The hydrochloric acid evolved is very corrosive.

Applications:

PVC-U: window frames, waste water pipes, undercoating of cars, instrument boards, acid containers, sheeting.

PVC-S: pipes, toys, balls, films for roof sealing and swimming pools, imitation leather ("vinyl").

Thermal analysis of PVC:

DSC: Characterization via glass transition (approx. −40 to +90℃, depending on plasticizer content), thermal stability.

TGA: Thermal stability, degradation behavior, filler content.

TMA: Glass transition, expansion behavior, expansivity.

DMA: Modulus, viscoelastic and damping behavior, glass transition, gelation.

6.6 聚醋酸乙烯, PVAC Polyvinyl Acetate

PVAC 不用于制造塑料部件：它太软了。市场上，最重要的 PVAC 形式是水中乳状液。干燥时，分散形成对所有各种基层附着良好的薄膜。

应用：PVAC 是乳状液粘合剂和涂料中的乳剂。增塑剂将玻璃化转变温度降低约 40℃ 甚至更多。

PVAC 的热分析：

DSC：通过玻璃化转变（约 40℃）来表征，含增塑剂的要低得多。

TGA：热稳定性、降解行为、填料含量。

TMA：玻璃化转变、膨胀行为、热塑性流动（冷流动）。

DMA：模量、粘弹和阻尼性能、玻璃化转变。

PVAC is not used to make plastic parts: it is too soft. Commercially, the most important form of PVAC is as an emulsion in water. On drying, the dispersion forms a film that adheres well to all kinds of substrates.

Applications: PVAC is the binder in emulsion adhesives and paints. Plasticizers lower the glass transition temperature of about 40℃ still further.

Thermal analysis of PVAC:

DSC: Characterization via glass transition (about 40℃), with plasticizer much lower.

TGA: Thermal stability, degradation behavior, filler content.

TMA: Glass transitions, expansion behavior, thermoplastic flow (cold flow).

DMA: Modulus, viscoelastic and damping behavior, glass transition.

6.7 聚酰胺, PA Polyamide

聚酰胺是半结晶热塑性聚合物，具有出色的机械性能，如抗冲击性和耐磨损性。浸在水中或暴露于空气中的聚酰胺可吸收一些水分。由于熔融温度高，高温下的尺寸稳定性出色。聚酰胺易于改性（共聚、矿物填料、纤维增强）。

应用：齿轮、皮带轮、轴承、螺旋桨、风扇、汽车部件、水笼头和通过纺丝时取向而成为强度特别高的纤维。

PA 的热分析：

DSC：通过熔融峰温、结晶度来鉴别 PA6、66、10、11、12。

TGA：降解行为、填料含量。

TMA：膨胀和收缩行为、膨胀性。

DMA：模量、粘弹和阻尼性能、玻璃化转变。

Polyamides are semicrystalline thermoplastic polymers with excellent mechanical properties such as toughness and wear strength. Polyamides immersed in water or exposed to ambient air absorb several percent of moisture. Due to the high melting temperatures, dimensional stability at elevated temperatures is excellent. Polyamides are easy to modify (copolymerization, mineral fillers, fiber reinforcement).

Applications: gear wheels, belt pulleys, bearings, screws, fans, car parts, water taps, and fibers, which become especially strong through orientation on spinning.

Thermal analysis of PA:

DSC: Identification of PA6, 66, 10, 11, 12 via melting peak temperature, crystallinity.

TGA: Degradation behavior, filler content.

TMA: Expansion and shrink behavior, expansivity.

DMA: Modulus, viscoelastic and damping behavior, glass transition.

6.8 聚对苯二甲酸乙二醇酯，PET Polyethylene Terephthalate

聚对苯二甲酸乙二醇酯 PET 是半结晶热塑性聚合物，通过从熔融状态骤冷能冻结在无定形状态。PET 球晶的生长速率约是 PE 的 500 分之一。这个特性被应用于 PET 的冷结晶：在玻璃化转变温度之上加热形成无定形板材或薄膜，同时结晶。结晶提高热变形温度（失去尺寸稳定性的温度）达 100K 以上。用这个方法，冷结晶了的 PET 变为不透明。薄膜或纤维取向后尽管结晶度高但是透明的。

应用：自从 1950 年以来，PET 主要用于衣服、船帆、无纺材料的耐磨、耐候纤维（例如，商品名达纶、涤纶、特雷维拉）和高拉伸强度的尺寸稳定性薄膜（录音带和录像带、计算机磁盘、书写用途）。一些机械部件如齿轮、轴承，近年来使用的软饮料瓶，均用 PET 制造。

PET 的热分析：
DSC：通过玻璃化转变温度、冷结晶和熔融峰温来鉴别。结晶度和氧化起始温度也重要。
TGA：降解行为、热稳定性、填料含量。
TMA：玻璃化转变、冷结晶、膨胀行为、膨胀性、热塑性流动。
DMA：模量、粘弹和阻尼性能、玻璃化转变。

Polyethylene terephthalate, PET, is a semicrystalline thermoplastic polymer that can be frozen in the amorphous state by shock-cooling from the molten state. The rate of growth of PET spherulites is about 500 times lower than with PE. This characteristic is applied in the cold crystallization of PET: Amorphous sheets or films are thermoformed above the glass transition temperature and crystallize at the same time. The crystallization increases the heat distortion temperature (temperature at which the dimensional stability is lost) by more than 100 K. In this way cold crystallized PET becomes opaque. Oriented films and fibers are transparent despite their high crystallinity.

Applications: Since 1950, PET has been used mainly for wear- and weather-resistant fibers for clothing, sails, and non-woven material (trade names, e.g. Dacron, Terylene, Trevira) and for dimensionally stable films of high tensile strength (for audio and videotapes, computer disks and graphical purposes). Some technical parts, such as gear wheels, bearings and in the recent years soft drink bottles are made from PET.

Thermal analysis of PET:
DSC: Identification via glassm transition temper-ature, cold crystallization and melting peak temperature. Degree of crystallinity and the oxidation onset temperature are also important.
TGA: Degradation behavior, thermal stability, filler content.
TMA: Glass transition, cold crystallization, expansion behavior, expansivity, thermoplastic flow.
DMA: Modulus, viscoelastic and damping behavior, glass transition.

6.9 聚碳酸酯，PC Polycarbonate

聚碳酸酯 PC 是高强度和高抗冲强度的无定形透明材料，它具有高热形变温度。

PC 的应用：高温下使用的专用部件（例如消毒瓶）、光盘、安全玻璃、照

Polycarbonate, PC, is an amorphous transparent material of high strength and high impact strength. It has a high heat distortion temperature.

Applications of PC: technical parts that are exposed to elevated temperatures (e.g. sterilizable bottles), compact disks, safety

| 相机外壳、车灯反光镜、可移动家用嵌装玻璃。 | glass, housings of cameras, reflectors for car headlights, mobile home glazing. |

PC 的热分析：

Thermal analysis of PC:

- DSC：通过玻璃化转变温度来鉴别和表征；结晶度只能在一些薄膜中测到。
- TGA：降解行为、热稳定性、填料含量。
- TMA：玻璃化转变、膨胀行为、膨胀性、热塑性流动。
- DMA：模量、粘弹和阻尼性能、玻璃化转变。

- DSC: Identification and characterization via glass transition; crystallinity can only be detected in certain films.
- TGA: Degradation behavior, thermal stability, filler content.
- TMA: Glass transition, expansion behavior, expansivity, thermoplastic flow.
- DMA: Modulus, viscoelastic and damping behavior, glass transition.

6.10 聚甲醛，POM Polyoxymethylene

聚甲醛 POM 或聚乙醛是半结晶的。由于硬度和刚度高，POM 是精密机械（螺钉，齿轮，自动连接件，办公室、家用、纺机的机械受力部件）的出色制造材料

Polyoxymethylene, POM, or polyacetal is semicrystalline. Due to its high hardness and stiffness, POM is an excellent construction material for precision mechanics (screws, gear wheels, snap connections, mechanically stressed parts of office, household and textile machines).

POM 的热分析：

Thermal analysis of POM:

- DSC：通过熔融行为和结晶度来表征。
- TGA：降解行为、填料含量。
- TMA：膨胀行为、膨胀性。
- DMA：模量、粘弹和阻尼性能、玻璃化转变。

- DSC: Characterization via melting behavior and crystallinity.
- TGA: Degradation behavior, filler content.
- TMA: Expansion behavior, expansivity.
- DMA: Modulus, viscoelastic and damping behavior, glass transition.

6.11 聚四氟乙烯，PTFE Polytetrafluoroethylene

聚四氟乙烯 PTFE 是具有良好耐磨损性和低摩擦力的半结晶热塑性聚合物。此外，静态和滑动摩擦系数相同：没有粘滑运动。甚至更为重要的是它出色的化学性能、热性能和介电性能。PTFE 的应用从很低的温度高至约 250℃。

Polytetrafluoroethylene, PTFE, is a semicrystalline thermoplastic polymer with favorable wear resistance and low friction. In addition, the coefficients for static and sliding friction are equal: no stick slip. Even more important are its excellent chemical, thermal and dielectric properties. PTFE has applications from very low temperatures up to approx. 250℃.

PTFE 微晶在 19℃ 有一个互变性的固——固转变。

PTFE crystallites undergo an enantiotropic solid-solid transition at 19℃.

PTFE 的应用：密封件、活塞环、管材、轴承、防粘表面涂层。

Applications of PTFE: seals, piston rings, pipes, bearings, anti-adhesive surface coatings.

PTFE 的热分析：
DSC：通过互变性转变和熔融行为、结晶度来鉴别和表征。
TGA：降解行为、填料含量。
TMA：膨胀行为、膨胀性、互变性转变。
DMA：模量、粘弹和阻尼性能、玻璃化转变。

Thermal analysis of PTFE:
DSC: Identification and characterization via enantiotropic transition and melting behavior, crystallinity.
TGA: Degradation behavior, filler content.
TMA: Expansion behavior, expansivity, enantiotropic transition.
DMA: Modulus, viscoelastic and damping behavior, glass transition.

7 热塑性聚合物的应用
Application of Thermoplastic Polymers

7.1 聚乙烯测试　Measurements on polyethylene

用峰温表征聚乙烯　PE, Characterization by Peak Temperature

样品	PE-LD 和 PE-LLD 球状颗粒，PE-HD 薄膜	$(-CH_2-CH_2-)_n$
Sample	PE-LD and PE-LLD as pellets, PE-HD as a film.	
条件	测试仪器：DSC	**Measuring cell**：DSC
Conditions	坩埚：	**Pan**：
	40 μl 标准铝坩埚，盖密封	Aluminum standard 40μl, lid hermetically sealed
	样品制备：	**Sample preparation**：
	PE-LD：从球状颗粒切出 9.796mg 的圆片	PE-LD：Disk of 9.796mg cut from pellet
	PE-LLD：从球状颗粒切出 8.128 mg 的圆片	PE-LLD：Disk of 8.128 mg cut from pellet
	PE-HD：从薄膜冲出 10.314 mg 的圆片	PE-HD：Disk of 10.314 mg punched from film
	DSC 测试：	**DSC measurement**：
	以 10K/min 从 30℃加热至 180℃	Heating from 30℃ to 180℃ at 10 K/min
	气氛：静态空气	**Atmosphere**：Static air

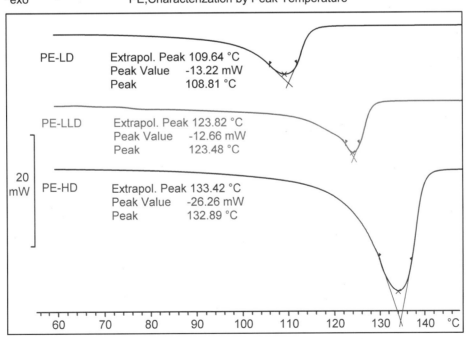

计算 Evaluation		PE-LD	PE-LLD	PE-HD
	外推峰温,℃ Extrapolated peak temperature in ℃	109.6	123.8	133.4
	峰温,℃ Peak temperature in ℃	108.8	123.5	132.9

注:两个值能用于表征峰温:拐点切线外推的交点和 DSC 曲线的最大值。外推峰温受随机峰形的影响较小,峰形可能由于随机噪音引起轮与轮的实际最大值的微小不同,导致"降温"的重现性不好。因此,外推值的重复性通常更好。

Note: Two values can be used to characterize the peak temperature: the extrapolated intersection point of the inflection tangents, and the maximum of the DSC curve. The extrapolated peak temperature is not influenced so much by the arbitrary shape of the peak where random noise may cause the actual maximum to differ slightly from run to run, leading to an irreproducible "peak temperature". For this reason the reproducibility of the extrapolated value is usually better.

解释 三种不同的 PE 具有不同典型的熔融峰,可利用各自降温加以鉴别。

Interpretation The melting peaks are typical for the three different PE qualities with their distinctive peak temperatures.

结论 DSC 熔融曲线的峰温是半结晶聚合物最常测定的性能。它是加工的关键参数,一般用于鉴别。

Conclusion The peak temperature of the DSC melting curve is the most frequently determined property of semicrystalline polymers. It is the key parameter for processing and serves generally for identification.

用结晶度表征聚乙烯　PE, Characterization by Crystallinity

样品　PE-LD 和 PE-LLD 球状颗粒，PE-HD 薄膜
Sample　PE-LD and PE-LLD as pellets，PE-HD as a film

条件　测试仪器：DSC　　　　　　　　　　　**Measuring cell**：DSC
Conditions　坩埚：　　　　　　　　　　　　　　**Pan**：
　　40μl 标准铝坩埚，盖密封　　　　　　　Aluminum standard 40μl, lid hermetically sealed
　　样品制备：　　　　　　　　　　　　　　　　**Sample preparation**：
　　PE-LD：从球状颗粒切出 9.796mg　　　PE-LD：Disk of 9.796 mg cut from pellet
　　　　　的圆片
　　PE-LLD：从球状颗粒切出 8.128mg　　PE-LLD：Disk of 8.128 mg cut from pellet
　　　　　的圆片
　　PE-HD：从薄膜冲出 10.314mg 的　　　PE-HD：Disk of 10.314 mg punched from film
　　　　　圆片
　　DSC 测试：　　　　　　　　　　　　　　　　**DSC measurement**：
　　以 10K/min 从 30℃加热至 180℃　　　Heating from 30℃ to 180℃ at 10 K/min
　　气氛：静态空气　　　　　　　　　　　　　　**Atmosphere**：Static air

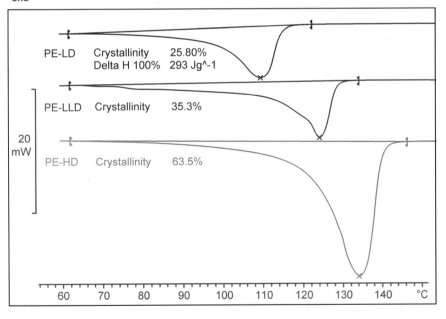

计算
Evaluation

	PE-LD	PE-LLD	PE-HD
结晶度，% Crystallinity in %	25.8	35.3	63.5
熔融热，% Heat of fusion in J/g	75.8	103.5	186.0

注：直线基线用于积分。由于使用归一化值，所以样品称量的准确很重要。

Note：A straight baseline has been used for integration. Since the normalized value is used, it is important that the sample weight is accurate.

解释 峰面积等于样品的熔融热。它与存在于样品中的微晶量成正比。结晶度由测得的熔融热与100%结晶聚乙烯的理论熔融热293J/g(见24页上的表)的定比来确定。

结论 DSC结晶度表示微晶相对于无定形材料的百分数。结晶度取决于分子结构的规整性如何(在有序间距内无支链或短支链)和热历史。由于硬度和强度随结晶度增大而增大,所以DSC结果与机械性能相关联。

Interpretation The area under the peak corresponds to the heat of fusion of the sample. It is proportional to amount of crystallites present in the sample. The degree of crystallinity is determined by comparing the measured heat of fusion with the theoretical heat of fusion of 100% crystalline polyethylene of 293 J/g (see table on page 24).

Conclusion The DSC crystallinity indicates the percentage of the material that is crystalline versus amorphous. The degree of crystallinity depends on how regular the structure of the molecule is (no branches or short branches in regular distance) and on the thermal history. Since hardness and strength increase with increasing crystallinity, the DSC result correlates with the mechanical properties.

| 用转化率曲线表征高密度聚乙烯 | PE-HD, Characterization by Conversion Curves |

样品 **Sample**	Lupolen 4261 球状颗粒和一个据称为 Lupolen 4261 的成型部件 Lupolen 4261 pellets and a molded part said to be Lupolen 4261	
条件 **Conditions**	测试仪器:DSC	**Measuring cell**:DSC
	坩埚: 40μl 标准铝坩埚,钻孔盖	**Pan**: Aluminum standard 40μl, pierced lid
	样品制备: 从球状颗粒或部件切出约 5mg 的圆片	**Sample preparation**: Disks of approx. 5.0 mg cut from pellet or part
	DSC 测试: 消除热历史:以 10K/min 从 25℃至 200℃,然后以 5K/min 冷却。 实际测试:以 10K/min 从 25℃至 200℃第二轮加热 气氛:氮气,50 cm³/min	**DSC measurement**: Elimination of thermal history:25℃ to 200℃ at 10 K/min, then cooling at 5 K/min. Actual measurement:second heating run from 25℃ to 200℃ at 10 K/min **Atmosphere**:Nitrogen, 50 cm³/min

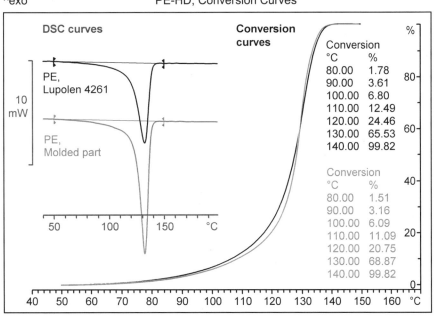

计算 **Evaluation**			Lupolen 4261	成型部件 Molded part
	熔融热,J/g Heat of fusion in J/g		152.0	177.0
	峰温,℃ Peak temperature in ℃		131.0	130.0
	峰宽,K Peak width in K		10.9	8.2
	熔融分数 Fraction Melted at	80℃,%	1.8	1.6
		100℃,%	6.8	6.0
		120℃,%	24.5	20.7
		140℃,%	99.8	99.8

解释 熔融行为用熔融热、峰温和半高的峰宽（3个值一起）或转化率曲线来表征。转化率曲线，也称为熔融分数，是在特定温度点晶区百分数的量度。它可用表格或图示的形式表示。与结晶度对比，转化率是相对值，因而实际上与称量的误差无关。用于本测试的天平的分辨率为0.1mg。例如0.3mg的误差会导致熔融热和结晶度计算值6%的误差。

结论 两个样品的熔融行为是相似的。主要不同在于熔融热。样品结晶行为的比较见下页。

Interpretation The melting behavior is characterized by the heat of fusion, the peak temperature and the peak width at half height (all 3 values together) or by the conversion curve. The conversion curve, also called the fraction melted, is a measure of the percentage of crystallites that has melted at the particular temperature. It can be presented in a tabular or graphical form. In contrast to the degree of crystallinity, conversion is a relative value and is therefore practically independent of weighing uncertainties. The balance used for this work had a resolution of 0.1 mg. An uncertainty of 0.3 mg can for example lead to a 6% error in the calculated value of the heat of fusion and hence in the degree of crystallinity

Conclusion The melting behavior of the two samples is similar. The main difference is seen in the heat of melting. The crystallization behavior of the samples is compared on the next page.

用结晶行为表征高密度聚乙烯
PE-HD, Characterization by Crystallization Behavior

样品 **Sample**	Lupolen 4261 球状颗粒和一个据称为 Lupolen 4261 的成型部件 Lupolen 4261 pellets and a molded part said to be Lupolen 4261		
条件 **Conditions**	测试仪器：DSC	**Measuring cell**：DSC	
	坩埚： 40μl 标准铝坩埚,钻孔盖	**Pan**： Aluminum standard 40μl, pierced lid	
	样品制备： 从球状颗粒或部件切出约 5mg 的圆片	**Sample preparation**： Disks of approx. 5.0 mg cut from pellet or part	
	DSC 测试： 以 10K/min 加热至 200℃熔融样品。 实际测试：以 10K/min 从 200℃冷却至 25℃	**DSC measurement**： Melting the sample by heating to 200℃ at 10K/min. Actual measurement：cooling from 200℃ to 25℃ at 10K/min	
	气氛：氮气,50 cm^3/min	**Atmosphere**：Nitrogen, 50 cm^3/min	

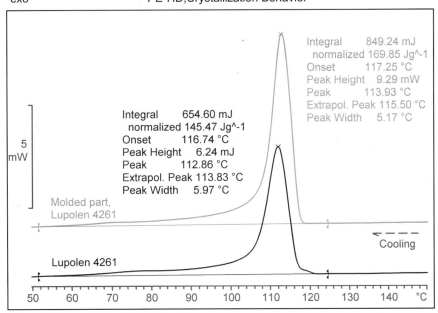

	Lupolen 4261	成型部件 Molded Part
结晶起始点,℃ Onset of crystallization in ℃	116.8	117.4
峰温,℃ Peak temperature in ℃	112.9	113.9
峰宽,℃ Peak width in ℃	5.9	5.0
结晶热,J/g Heat of crystallization in J/g	145.0	170.0

计算 / Evaluation

解释 结晶行为用结晶起始点、峰宽和结晶热来表征。此外,转化率曲线和表格在比较用途方面较为有用。

结论 两个样品的结晶行为是相似的。
在前一个实验中比较了相同样品的熔融行为。结晶热与前面熔融实验中测得的差异大约等量。

Interpretation The crystallization behavior is characterized by the crystallization onset, the peak width and the heat of crystallization. In addition, the conversion curve and table are useful for comparison purposes.

Conclusion The crystallization behavior of the two samples is similar.
The melting behavior of the same samples is compared in the previous experiment. The heats of crystallization differ by about the same amount as measured before in the melting experiment.

来自不同制造商的高密度聚乙烯 PE-HD from Different Manufacturers

样品	Vestolen A 6017, Stamylan HD 7771, NCPE 2233, Finathene 58070, Lupolen 4261	
Sample	Vestolen A 6017, Stamylan HD 7771, NCPE 2233, Finathene 58070, Lupolen 4261	
条件	测试仪器：DSC	**Measuring cell**：DSC
Conditions	坩埚：	**Pan**：
	40 μl 标准铝坩埚，钻孔盖	Aluminum standard 40 μl, pierced lid
	样品制备：	**Sample preparation**：
	从球状颗粒切出的圆片	Disk cut from pellet
	DSC 测试：	**DSC measurement**：
	以 10 K/min 从 30 ℃ 加热至 200 ℃	Heating from 30 ℃ to 200 ℃ at 10 K/min
	气氛：氮气，50 cm^3/min	**Atmosphere**：Nitrogen, 50 cm^3/min

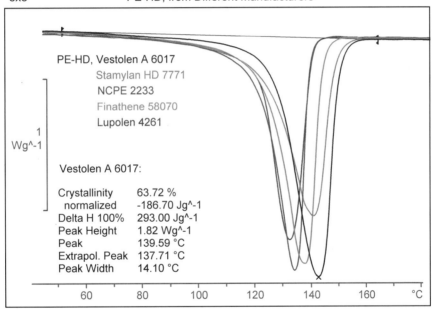

	名称 Name	结晶度，% Crystallinity in %	峰温，℃ Peak temperature in ℃	峰宽，℃ Peak width In ℃	样品质量，mg Sample mass in mg
计算 **Evaluation**	Vestolen A 6017	63.7	139.6	14.1	13.3
	Stamylan HD 7771	55.7	136.6	15.2	22.5
	NCPE 2233	52.4	133.1	10.6	4.9
	Finathene 58070	58.4	135.2	12.4	11.3
	Lupolen 4261	48.2	131.4	11.5	4.5

注：为了简洁，图中只对 Vestolen 的曲线进行了计算。
Note：For clearness, only the Vestolen curve is evaluated in the diagram.

解释 所有样品的 DSC 曲线在 50℃和 60℃之间开始偏离基线。信号在 145℃和 160℃之间回到基线。

结论 不同种类的 PE-HD 的 DSC 熔融峰显示不同的特征。为了消除由于各自的热历史产生的影响，样品应该以 10K/min 预熔和再结晶。此外，样品质量应该一致，比如 5±2mg。特别是峰宽依赖于样品质量（增大样品质量→增加熔融时间→峰宽度增加）。

Interpretation The DSC curves of all samples begin to deviate from baseline between 50℃ and 60℃. The signal returns to the baseline between 145℃ and 160℃.

Conclusion The DSC melting peaks of PE-HD of different quality show characteristic differences. To eliminate effects due to individual thermal history, the samples should have been pre-melted and recrystallized at 10 K/min. In addition, the sample masses should be more uniform, e.g. 5±2 mg. The peak width in particular depends on the sample mass (increasing mass → increasing melting time → broader peak).

聚乙烯的熔融曲线和热历史　　PE, Melting Curve and Thermal History

样品 **Sample**	PE-HD 薄膜 PE-HD film	
条件 **Conditions**	测试仪器：DSC 坩埚： 40μl 标准铝坩埚，盖密封 样品制备： 从薄膜冲出 2.33mg 的圆片 DSC 测试： 预处理：在 129℃下退火 60 分钟，接着以 5K/min 冷却至 40℃，以 5K/min 从 40℃加热至 180℃（见 129℃下退火的 PE-HD 曲线），然后以 5K/min 从 160℃冷却至 40℃ 第二轮加热以 5K/min 从 40℃至 180℃（见消除了热历史的 PE-HD 曲线） 气氛：静态空气	Measuring cell：DSC Pan： Aluminum standard 40μl, lid is hermetically sealed Sample preparation： Disk of 2.33 mg punched out of film DSC measurement： Pretreatment：annealing for 60 min at 129℃ followed by cooling to 40℃ at 5 K/min, Heating from 40℃ to 180℃ at 5 K/min (see curve of PE-HD annealed at 129℃), then cooling from 160℃ to 40℃ at 5 K/min Second heating run from 40℃ to 180℃ at 5 K/min (see curve of PE-HD with thermal history eliminated) Atmosphere：Static air

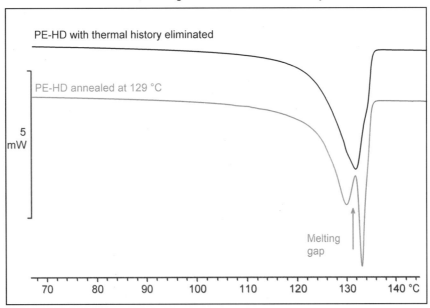

计算　熔融中断处可计算为起始点，或用转化率曲线来表征熔融行为。然而，它所展现的最重要的信息是在退火温度下进行。

Evaluation　The melting gap can be evaluated as an onset, or the melting behavior can be characterized by the conversion curve. However, the most important information it reveals is the temperature at which annealing was performed.

解释 微晶分离发生在129℃下退火期间：小微晶熔化，重新结晶形成熔点高于129℃的微晶。在该过程中，当然不能形成熔点等于或低于129℃的微晶。当样品冷却时，会再形成较低熔点的小微晶，但在样品的第二轮加热中，在129℃处出现了中断（称为熔融中断）。熔融完成后，所用热历史被消除。

结论 DSC熔融曲线的形状取决于样品的热历史。熔融中断常用来检查加工温度是否正确，例如，PE高压电缆的退火。完全的熔融消除了所有的热历史，这是比较不同原材料的先决条件。

Interpretation Crystallite segregation occurs during annealing at 129 ℃： Small crystallites melt and recrystallize to form crystallites with melting points above 129℃. Crystallites with melting points of 129℃ or lower cannot of course be formed during this process. As the sample is cooled, small crystallites with lower melting points form again, but a gap at 129℃ (called the melting gap) appears in the second heating run of the sample. On complete melting, any thermal history is eliminated.

Conclusion The shape of the DSC melting curve depends on the thermal history of the sample. The melting gap is often used to check that the processing temperature was correct, e. g. the annealing of PE high voltage cables. Complete melting eliminates any thermal history and is a prerequisite for the comparison of different raw materials.

高密度聚乙烯电缆管线和再生料的鉴定
PE-HD, Identification of Cable Tubing and Recycled Material

样品 **Sample**	两种不同的汽车用管材和据称为 PE-HD 的再生料 Two different types of tubing for automobiles and recycled material said to be PE-HD		
条件 **Conditions**	测试仪器：DSC 坩埚： 40μl 标准铝坩埚,钻孔盖 样品制备： 从材料切出的圆片 DSC 测试： 以 10K/min 从 30℃ 加热至 250℃ 气氛：氮气,50 cm³/min	Measuring cell：DSC Pan： Aluminum standard 40μl, pierced lid Sample preparation： Disk cut from material DSC measurement： Heating from 30℃ to 250℃ at 10 K/min Atmosphere：Nitrogen, 50 cm³/min	

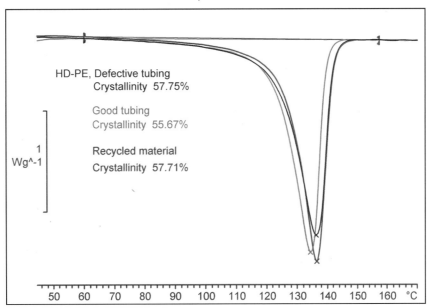

计算 **Evaluation**		结晶度,% Crystallinity in %	峰温,℃ Peak temperature in ℃
	有缺陷的管材 Defective tubing	57.8	135.2
	好的管材 Good tubing	55.7	133.2
	再生料 Recycled material	57.7	134.8

解释 三个样品的 DSC 熔融曲线均为 PE-HD 特有的。

结论 DSC 熔融曲线能清楚地鉴定 PE-HD。有缺陷的管形材料也是由正确的材料制造的。为了消除热历史的效应,样品应该预熔,然后以同样的速率冷却,例如 10 K/min。再生料没有用正确标签的原制造商包装材料发货,因此,确认聚合物是正确的材料非常重要。还必须检查再生料以确认最初的处理和再生产加工不会显著降低质量。

Interpretation The DSC melting curves of all three samples are characteristic for PE-HD.

Conclusion The DSC melting curve allows PE-HD to be clearly identified. The defective tubing is also made of the correct material. To eliminate the effects of thermal history, the samples should be premelted and then cooled at identical rates, e.g. 10 K/min.

Recycled material is not delivered in properly labeled original manufacturer's packaging material. It is therefore very important to confirm that the polymer is the right material. The recycled material must also be checked to make sure that the initial processing and the recycling did not degrade the quality to any significant extent.

高密度聚乙烯再生板的 DSC   DSC of Recycled Sheets, Said to be PE-HD

样品 明显发脆和变形的板材,据称由 PE-HD 制造
Sample Noticeably brittle and distorted sheets, said to be made of PE-HD

条件 测试仪器:DSC   **Measuring cell**:DSC
Conditions 坩埚: **Pan**:
40μl 标准铝坩埚,钻孔盖   Aluminum standard 40μl, pierced lid
样品制备: Sample preparation:
从材料切出的圆片(6.8mg)   Disk cut from material (6.8 mg)
DSC 测试: DSC measurement:
以 10K/min 从 30℃加热至 200℃   Heating from 30℃ to 200℃ at 10K/min
气氛:氮气,50cm³/min   Atmosphere: Nitrogen, 50cm³/min

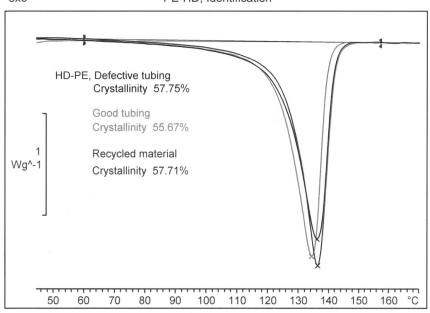

计算 测得的 PP 峰熔融热(14.3J/g)能大致计算 PP 的含量。如果取纯 PP 的熔融热为 60J/g(聚丙烯的典型值),则含量由下式给出:

含量 = $\Delta H/\Delta H_{pp}$ · 100%,即 23.9%

Evaluation The heat of fusion determined for the PP peak (14.3 J/g) allows the approximate content of PP to be calculated. If the heat of fusion of pure PP is taken to be 60 J/g (typical for polypropylene), then the content is given by:

Content = $\Delta H/\Delta H_{PP}$ · 100%, i.e. 23.9%

解释 熔融曲线清楚地显示该材料是 PE-HD(峰温 135℃)和 PP(峰温 163℃)的共混物,不是纯 PE。

Interpretation The melting curve clearly shows that the material is a blend of PE-HD (peak temperature of 135℃) and PP (peak temperature of 163℃) and not pure PE.

结论 DSC 证明该材料不符合质量的要求。因此,供应商免费更换了板材。

Conclusion DSC proves that the material does not match the quality agreed upon. The supplier therefore replaced the sheets free of charge.

低密度聚乙烯的两个产品的比较 PE-LD, Comparison of Two Products

样品	Lupolen 1800 S 和 Lupolen conz. SL 020 A 球形颗粒	
Sample	Lupolen 1800 S and Lupolen conz. SL 020 A as pellets	
条件	测试仪器:DSC	**Measuring cell**:DSC
Conditions	坩埚:	**Pan**:
	40μl 标准铝坩埚,钻孔盖	Aluminum standard 40μl, pierced lid
	样品制备:	**Sample preparation**:
	从球形颗粒中心部分切出的圆片	Disks cut from center part of pellet
	DSC 测试:	**DSC measurement**:
	以 10K/min 从 30℃加热至 250℃	Heating from 30℃ to 250℃ at 10 K/min
	气氛:空气,50cm³/min	**Atmosphere**:Air, 50cm³/min

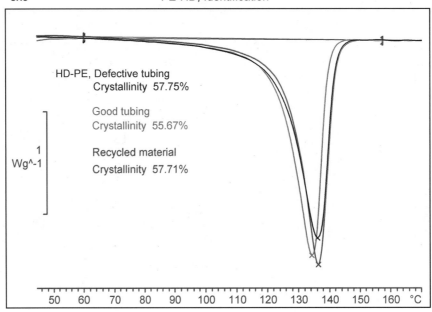

计算 / Evaluation

	Lupolen 1800 S	Lupolen conz. SL 020 A
结晶度,% Crystallinity in %	29.2	35.6
峰温,℃ Peak temperature in ℃	108.9	114.9
氧化起始点,℃ Onset of oxidation in ℃	220.0	217.0
样品质量,mg Sample mass in mg	19.5	14.35

解释 DSC 曲线具有 PE-LD 的典型形状。200℃以上的放热反应是由于开始氧化产生的。

结论 结晶度和峰温的计算显示这两个样品具有类似的特征，都是 PE-LD。大约 220℃的氧化起始点显示它们被"稳定化处理"过，因为起始点稍稍高于对基本 PE 的预期。通过添加另外的抗氧剂可进一步稳定化，这可将起始点移动到高至 260℃。

Interpretation The DSC melting curves have the typical shape for PE-LD. The exothermic reaction above 200℃ is due to the onset of oxidation.

Conclusion Evaluation of the percentage crystallinity and peak temperature reveal that these two samples are of similar quality and that both are PE-LD. The onsets of oxidation around 220℃ indicate they are "processing stabilized" because the onsets are slightly higher than that expected for basic PE. Lupolen 1800 S is more stable than the other sample. Further stabilization could be achieved by adding additional antioxidants. This could shift the onset to as high as 260℃.

聚乙烯的氧化稳定性 PE, Oxidation Stability

样品 **Sample**	Lupolen GM5040 T12（稳定化的 PE-HD 薄膜） Lupolen GM5040 T12 (stabilized PE-HD film)	
条件 **Conditions**	测试仪器：DSC	**Measuring cell**：DSC
	坩埚： 40μl 标准铝坩埚，不加盖 40μl 标准铜坩埚，不加盖，作比较	Pan： Aluminum standard 40μl without lid, 40μl copper pan without lid, for comparison
	样品制备： 从薄膜冲出的圆片	Sample preparation： Disk punched out of film
	DSC 测试： 按照 EN 728（欧洲标准）： 在氮气下，50 cm^3/min 在 30℃下恒温 3 min，然后以 20K/min 加热至 200℃，在 200℃恒温 2 min。 在氧气下，50 cm^3/min（由气体控制器自动切换）： 在 200℃下恒温开始实际测试。由于只需要氧化起始点，可使用条件实验终止软件选项缩短测试时间。	DSC measurement： According EN 728 (European standard)： Under nitrogen, 50 cm^3/min 3 min isothermal at 30℃, then heating to 200℃ at 20 K/min, 2 min isothermal at 200℃. Under oxygen, 50 cm^3/min (automatically switched by the gas controller)： Isothermally at 200℃ starts the actual measurement. Since only the onset of oxidation is required, the actual measurement time can be reduced using the Conditional Experiment Termination software option.
	气氛： 分别为氮气和氧气，由气体控制器自动切换	Atmosphere： Nitrogen and oxygen, respectively, automatically switched by the gas controller

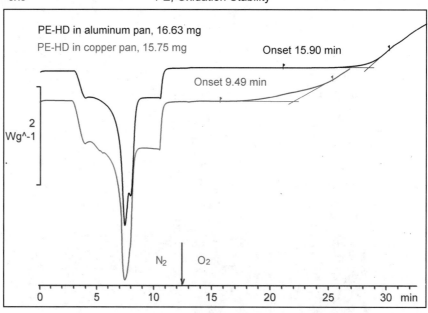

计算 在铝坩埚中的氧化诱导时间（OIT）是 15.9 min，在铜坩埚中的是 9.5 min。熔融曲线也可计算。

解释 切换到氧气后至氧化起始点（基线与切线的交点）的时间称为诱导期或氧化诱导时间 OIT。OIT 长显示稳定性好，反之亦然。与铜或铜合金接触催化氧化反应。受铜影响的聚烯烃需要专门的添加剂来中和催化效应，这种添加剂的功效通过样品在铜坩埚中进行的实验来测定。好的"铜稳定性"产生的 OIT 至少是在铝坩埚中得到值的一半。没有这种添加剂，在铜坩埚中的 OIT 约为十分之一。

注：聚烯烃给出明确的诱导期。其它聚合物加热时最好以较低的速率在氧气中测试，5K/min 较适宜，直到能看到清晰的放热氧化。

结论 用 DSC 测定氧化稳定性既快速又容易。特别建议给长期使用的物品如电缆和水管作质量保证用。应该测试每一批原材料。此外，从测得的熔化峰能鉴定聚合物。

Evaluation The oxidation induction time (OIT) in the aluminum pan is 15.9 min, in the copper pan 9.5 min. The melting curve could also be evaluated.

Interpretation The time after switching to oxygen to the onset of oxidation (intersection of baseline with the tangent) is called the induction period or Oxidation Induction Time, OIT. A long OIT indicates good stability and vice versa. Contact with copper or copper alloys catalyzes the oxidation reaction. Polyolefins that will be exposed to copper require special additives that neutralize this catalytic effect. The efficiency of such additives is determined by carrying out the experiment with the sample in a copper pan. Good "copper stability" results in an OIT of at least half the value obtained in the aluminum pan. Without such additives, the OIT is about 10 times shorter in copper pan.

Note: Polyolefines give distinct induction periods. Other polymers are best tested by exposure to oxygen on heating at relatively low rates, preferably 5 K/min, until distinct exothermal oxidation is visible.

Conclusion The determination of the oxidative stability by DSC is fast and easy. It is especially recommended for the quality assurance of long-life goods such as electrical cables and water pipes. Each batch of the raw material should be measured. In addition, the polymer can be identified from the fusion peak obtained.

用动态负载 TMA 测试交联聚乙烯　　Cross-Linked PE by Dynamic Load TMA

样品
Sample
交联聚乙烯板材
Cross-linked polyethylene sheet

条件
Conditions

测试仪器：DSC
坩埚：
在直径 6 mm 和厚度 0.5 mm 的石英圆片之间测试样品

样品制备：
用刀片切出约 4×4 mm 的矩形样品,样品高度 5.1 mm,与原板材的厚度一样。

TMA 测试：
第一轮加热以 10 K/min 至 150 ℃ 以消除热历史。自然冷却至 25 ℃ 后,以 10 K/min 的加热速率进行测试。
注:预处理也是在动态负载下完成的。

负载：
在 0.01 和 0.19N 之间每 6s 周期性变化

气氛:静态空气

Measuring cell：DSC
Pan：
The sample is measured between fused silica disks of 6mm diameter and 0.5mm thickness

Sample preparation：
A rectangular sample of about 4x4 mm is cut off with a knife. The height of the sample, 5.1 mm, is the same as the thickness of the original sheet

TMA measurement：
A first heating run to 150℃ at 10 K/min eliminated the thermal history. After uncontrolled cooling to 25℃, the measurement was performed at a heating rate of 10 K/min.
Note: the pretreatment was also done under dynamic load.

Load：
Periodically changing every 6 s between 0.01 and 0.19 N

Atmosphere：Static air

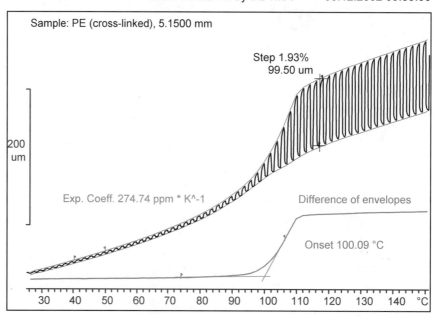

计算 无定形状态 1.9% 弹性形变处的复合杨氏模量 E 可按如下计算：

$$E=\Delta F/(A \cdot \Delta L_r),$$

式中 ΔF 是力的变化，A 是样品的截面积，ΔL_r 是相对长度变化(1.9%)。

$$E=0.18\,\text{N}/(4 \cdot 4\,\text{mm}^2 \cdot 0.019)=0.59\,\text{N/mm}^2$$

100.09℃时包络线之差的起始温度与微晶的熔融有关。

从上包络线（负载 0.01N）的斜率得到的膨胀系数为：

40℃和50℃之间，半结晶 275 ppm/K
120℃和130℃之间，无定形 353 ppm/K

解释 在微晶熔程以下的交联聚乙烯是坚硬的，就像标准PE。当微晶熔融时，体积增加，性能类似橡胶态。有一个由变化的力产生的约1.9%的弹性形变。上下包络线之间的距离与柔量(1/E)成正比。曲线包络的总斜率反映了样品的膨胀。

注：不交联的PE会在熔融后被挤压出圆片间（塑性形变）。

结论 在微晶熔程以上交联聚乙烯的热机械性能与标准PE的完全不同。不是粘性流动，而是有橡胶态的弹性行为。交联大分子阻止了塑性形变。

Evaluation The elastic deformation of 1.9% in the amorphous state allows the complex Young's modulus, E, to be calculated as follows:

$$E=\Delta F/(A \cdot \Delta L_r),$$

where ΔF is the change in force, A the cross-sectional area of the sample, and ΔL_r the relative change in length (1.9%).

$$E=0.18\,\text{N}/(4 \cdot 4\,\text{mm}^2 \cdot 0.019)=0.59\,\text{N/mm}^2$$

The onset of the difference of the envelopes at 100.09℃ is related to the crystallite melting.

The expansion coefficients derived from the slope of the upper envelope (load 0.01 N) are:

between 40℃ and 50℃, semicrystalline 275 ppm/K
between 120℃ and 130℃, amorphous 353 ppm/K

Interpretation Cross-linked polyethylene below the crystallite melting range is rigid, just like standard PE. When the crystallites melt the volume increases and the properties become rubbery-like. There is an elastic deformation of approx. 1.9% caused by the varying force. The distance between upper and lower envelopes is proportional to the compliance (1/E). The overall slope of the curve envelope reflects the expansion of the sample.

Note: PE that is not cross-linked would be squeezed out between the disks after melting (plastic deformation).

Conclusion The thermomechanical properties of cross-linked polyethylene above the crystallite melting range are completely different to those of normal PE. Instead of viscous flow there is rubbery elastic behavior. The cross-linked macromolecules prevent plastic deformation.

聚乙烯的 TGA 成分分析　　PE, Compositional Analysis by TGA

样品	PE-LD，Lacqtène 1002TN22		
Sample	PE-LD，Lacqtène 1002TN22		

条件	测试仪器：DSC	Measuring cell：DSC
Conditions	坩埚：	Pan：
	70μl 氧化铝坩埚，不带盖	Alumina 70μl, no lid
	样品制备：	Sample preparation：
	从球状颗粒切出 6.0570 mg	6.0570 mg cut from a pellet
	TGA 测试：	TGA measurement：
	以 30 K/min 从 40℃ 开始加热，在 650℃ 自动气体切换	Heating from 40℃ at 30 K/min with automatic gas switching at 650℃
	气氛：	Atmosphere：
	氮气，50cm^3/min；650℃ 以上为空气，50cm^3/min	Nitrogen，50cm^3/min；above 650℃ air，50cm^3/min

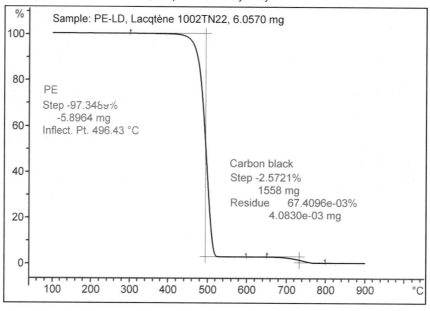

计算　对两个台阶都进行水平切线计算：

PE 含量	97.3%
TGA 拐点	496 ℃
碳黑含量	2.57%

系列实验显示，取自同一颗粒的 4 个样品，碳黑含量的标准偏差为 0.014%。但取自几个不同颗粒时标准偏差为 0.25%。

Evaluation　Both steps were evaluated with horizontal tangents.

PE content	97.3%
TGA inflection point	496℃
Carbon black content	2.57%

A series of measurements showed a standard deviation for the carbon black content of 0.014% when the 4 samples originated from one pellet, but 0.25% when several different pellets were used.

解释 在氮气下主要成分聚乙烯在400℃至600℃之间热解。余下的碳黑在切换气氛至空气后燃烧。碳黑的比表面积即"活性"越大,燃烧得越快。

注:在PE测试前先要做一个空白实验。空白曲线会在随后的测试中自动被扣除。

结论 用TGA进行成分分析既快速又准确。此外,还可获得表征PE的定性信息(TGA 拐点)。测试几个取自不同颗粒的样品可评估产品的一致性。

Interpretation Polyethylene, the main component, pyrolyzes between 400℃ and 600℃ under nitrogen. The remaining carbon black burns after the switching the atmosphere to air. The higher the specific surface area or "activity" of the carbon black, the faster the combustion.

Note: A blank experiment was run prior to the PE measurement. The blank curve was automatically subtracted from subsequent measurements.

Conclusion Compositional analysis by TGA is fast and accurate. In addition, qualitative information is obtained that characterizes the PE (TGA inflection point). Measuring samples from several different pellets allows the homogeneity of the product to be assessed.

高取向高密度聚乙烯纤维的 DSC   DSC of Highly Oriented PE-HD Fiber

样品 **Sample**	Dyneema 纤维，由 DSM 通过凝胶纺丝加工生产 Dyneema fiber, produced by the DSM gel spinning process	
条件 **Conditions**	测试仪器：DSC	**Measuring cell**：DSC
	坩埚：	**Pan**：
	40μl 标准铝坩埚，密封，为了确保坩埚与样品的良好热接触，在密封前对盖子进行凹槽压制。	Aluminum standard 40μl, hermetically sealed, lid indented prior to sealing in order to ensure good thermal contact between pan and sample.
	样品制备：	**Sample preparation**：
	将长度约为 25 cm (1.235 mg) 的 PE 纤维卷成一个小球放进坩埚中	A length of approx. 25 cm (1.235 mg) of the PE fiber was rolled to a small ball and placed in the pan
	DSC 测试：	**DSC measurement**：
	所有加热都以 10 K/min 从 30℃ 至 180℃。第二轮加热前骤冷样品。第三轮加热前骤冷样品，然后在 30℃ 放置 10 小时。第三轮加热后，以 0.5 K/min 冷却至 100℃。骤冷在自动进样器上进行。	All heating runs were from 30℃ to 180℃ at 10 K/min. Before the second run, the sample was quench-cooled. Before the third run, it was quench-cooled and then stored at 30℃ for 10 h. After the third run, it was cooled to 100℃ at 0.5 K/min. Quench-cooling was performed in the sample changer
	气氛：静态空气	**Atmosphere**：Static air

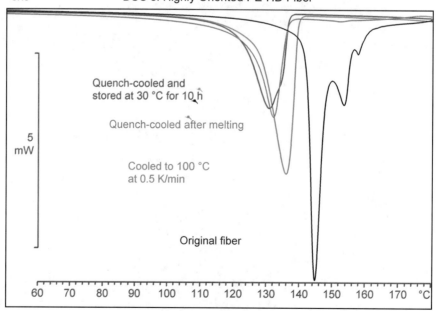

计算 对直线基线上的峰积分得到：
Evaluation Integrating the peaks over a straight baseline gives：

热历史 Thermal history	峰温,℃ Peak temp. in ℃	熔融热,J/g Heat of fusion in J/g
原始凝胶纺丝 Original gel spinning	144	253
骤冷 Quench-cooled	132	125
骤冷并在 30℃下放置 10 小时 Quench-cooled and stored at 30℃ for 10h	131	145
以 0.5 K/min 冷却至 100℃ Cooled to 100℃ at 0.5 K/min	136	186

解释 所有加热曲线都显示因聚合物结晶区域熔融产生的吸热效应。第一个曲线的熔融峰与后面曲线的熔融峰完全不同。Dyneema 纤维是由通过冷却拉伸而获得最大取向的特殊高分子量 PE 制成。这种高规整度使得熔程移至较高温度。在熔融时，它被不可逆地破坏。

Interpretation All the heating curves show endothermic effects caused by the fusion of the crystalline regions of the polymer. The first fusion peak is completely different to the curves that follow. The Dyneema fiber is made of especially high molecular mass PE that is stretched on cooling to obtain maximum orientation. This high degree of order shifts the melting range to much higher temperatures. It is irreversibly destroyed on melting.

结论 DSC 用来检测和比较半结晶聚合物的规整度。破坏高有序度是很容易的，但是重建该有序度实际上不可能。

可在自动进样器上自动完成骤冷样品：在测试结束后将热坩埚放在转盘上，它会在几秒钟内冷却至室温。然后可对骤冷样品进行下一轮的测试。

Conclusion DSC is used to detect and compare the degree of order in semicrystalline polymers. It is easy to destroy a high degree of order. Reconstruction of the order is virtually impossible.

Quench-cooling a sample is done automatically in the sample changer：at the end of a measurement the hot pan is placed on the cold turntable where it cools to ambient temperature within a few seconds. The next measurement can then be performed with the quench-cooled sample.

7.2 聚丙烯测试 Measurements on polypropylene based material

样品质量对聚丙烯的影响 PP, Influence of The Sample Mass

样品 **Sample**	Hostalen PPN2060 球状颗粒 Hostalen PPN2060 as pellets	
条件 **Conditions**	测试仪器：DSC 坩埚： 40μl 标准铝坩埚，密封 样品制备： 从球状颗粒中心部分切出圆片 DSC 测试： 以 5 K/min 在 40℃ 与 200℃ 之间加热和冷却。 气氛：静态空气	Measuring cell：DSC Pan： Aluminum standard 40μl, hermetically sealed Sample preparation： Disk cut from center part of pellet DSC measurement： Heating and cooling between 40℃ and 200℃ at 5 K/min. Atmosphere：Static air

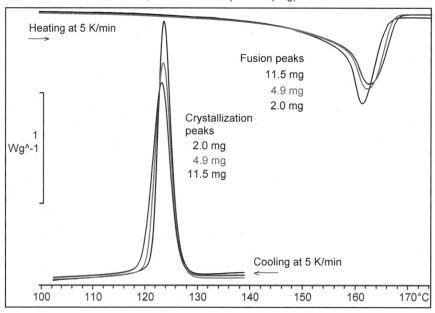

计算 Evaluation	样品质量,mg 加热 Sample mass in mg On Heating	2.0	4.9	11.5
	熔融热,J/g Heat of fusion in J/g	88.5	97.2	97.8
	峰高,mW Peak height in mW	1.6	3.3	7.2
	标准峰高,W/g Peak height normalized in W/g	0.8	0.7	0.6
	峰温,℃ Peak temperature in ℃	161.3	162.2	163.3
	峰宽,K Peak width in K	5.3	6.8	7.7

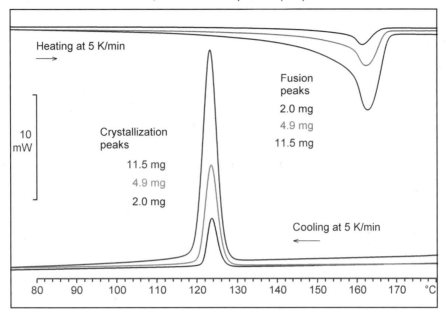

样品质量, mg 冷却 Sample mass in mg On Cooling	2.0	4.9	11.5
结晶热, J/g Heat of crystallization in J.g	91.5	93.1	92.9
峰高, mW Peak height in mW	4.6	9.5	20.0
标准峰高, W/g Peak height normalized in W/g	2.3	2.0	1.8
峰温, ℃ Peak temperature in ℃	123.8	123.9	124.0
峰宽, K Peak width in K	2.8	3.5	3.8

解释 正如预期的，峰高(mW)随着样品质量的增加而增加。由于大样品需要更多的时间来熔融，因此峰宽也增加。归一化峰高相互比较接近，当然半峰高的峰宽仍是一样的。

结论 样品质量略微影响了熔融与结晶行为。为了进行比较，质量应该相近，例如 5±2mg。随着质量的增加，热效应更持久，因为要传递更多的热量。特征温度也有偏移。归一化到用每克瓦单位表示后有助于补偿峰高的差异。

Interpretation As expected, the peak heights (in mW) increase with increasing mass. Since large samples require more time to melt, the peak width also increases. The normalized peak heights are closer together; the peak width at half height remains the same of course.

Conclusion The sample mass influences melting and crystallization behavior to a slight extent. For comparison purposes, the masses should therefore be similar, e.g. 5±2 mg. With increasing sample mass, the thermal effects last longer because greater quantities of heat have to be transferred. The characteristic temperatures also shift. Normalizing the presentation in watts per gram helps to compensate for the differences in peak height.

来自不同制造商的聚丙烯　PP from Different Manufacturers

样品 **Sample**	Ferrolene HU10TS3, Hostalen PPH 1050, Hostalen PPR 1042, Hostalen PPX 4305, Procom GC30H251, Vestolen P7032G, 都是球状颗粒 Ferrolene HU10TS3, Hostalen PPH 1050, Hostalen PPR 1042, Hostalen PPX 4305, Procom GC30H251, Vestolen P7032G, all as pellets	
条件 **Conditions**	测试仪器：DSC 坩埚： 40μl 标准铝坩埚，钻孔盖 样品制备： 从球状颗粒中心部分切出圆片 DSC 测试： 以 10 K/min 从 30℃加热至 250℃ 气氛：氮气, 50cm³/min	Measuring cell：DSC Pan： Aluminum standard 40μl, pierced lid Sample preparation： Disk cut from center part of pellet DSC measurement： Heating from 30℃ to 250℃ at 10 K/min Atmosphere：Nitrogen, 50cm³/min

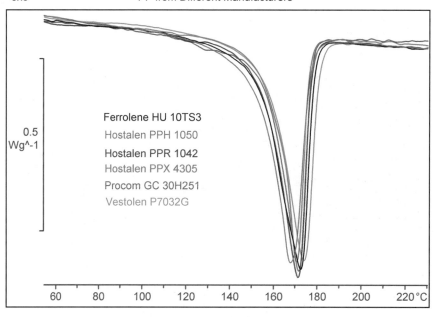

计算 / Evaluation

名称 Name	样品质量, mg Sample mass in mg	结晶度, % Crystallinity in %	PP 峰温, ℃ PP Peak Temp. in ℃	PE 峰温, ℃ PE Peak Temp. in ℃
Ferrolene, HU10TS3	9.5	36.4	171.5	—
Hostalen PPH 1050	12.4	36.4	166.5	—
Hostalen PPR 1042	9.2	34.7	170.1	124.0
Hostalen PPX 4305	11.1	27.3	169.4	—
Procom GC30H251	5.6	34.0	170.9	128.0
Vestolen P7032G	11.3	33.1	172.7	—

解释 PP 均聚物的 DSC 熔融曲线通常只有单个熔融峰,而与乙烯形成的 PP 共聚物常在 125℃附近还有一个附加的小峰或肩。在测定 PP 结晶度时,PE 峰也计算了进去。

结论 聚丙烯可很容易的用 DSC 进行表征。

Interpretation The DSC melting curves of PP homopolymers usually consist of a single melting peak, while PP copolymers with ethylene often have an additional small peak or a shoulder around 125℃. For the determination of PP crystallinity, the PE peak is also included in this evaluation.

Conclusion Polypropylene can easily be characterized by DSC.

空气中聚丙烯的 DSC 测试 PP, DSC Measurements in Air

样品 **Sample**	Hostalen PPN 1034/12 球状颗粒 Hostalen PPN 1034/12 as pellet		
条件 **Conditions**	测试仪器:DSC	**Measuring cell**:DSC	
	坩埚: 40 μl 标准铝坩埚,钻孔盖	**Pan**: Aluminum standard 40 μl, pierced lid	
	样品制备: 从球状颗粒中心部分切下圆片	**Sample preparation**: Disk cut from center part of pellet	
	DSC 测试: 以 10 K/min 从 30℃加热至 300℃	**DSC measurement**: Heating from 30℃ to 300℃ at 10 K/min	
	气氛:空气,50 cm³/min	**Atmosphere**: Air, 50 cm³/min	

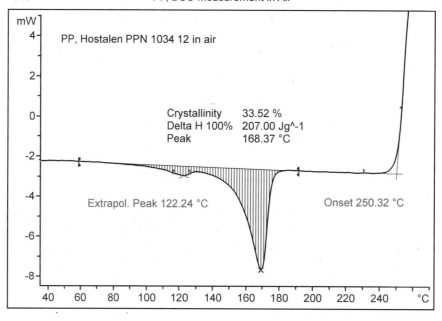

计算 **Evaluation**		
	结晶度,% Crystallinity in %	33.5*
	PP 峰温,℃ Peak temperature PP in ℃	168.4
	PE 峰温,℃ Peak temperature PE in ℃	122.2
	氧化起始点,℃ Onset of oxidation in ℃	250.3

*积分:60 至 193℃。100%结晶 PP 的熔融热取 207 J/g。
*Integration: 60 to 193℃. The heat of fusion of 100% crystalline PP is taken to be 207 J/g.

解释 约122℃的小峰表明PP共聚物含有乙烯。氧化分解在远高于PP熔融峰后出现,这表明存在于空气中的氧不影响熔融曲线。尽管如此,一些标准还是要求在氮气下测试熔融峰。

结论 如果没有自动气体控制器,也可在空气下测试熔融峰。氧化反应的起始点是稳定性的量度(温度越高,稳定性越好)。应用实例"聚乙烯的氧化稳定性"讨论了等温氧化诱导时间OIT。

Interpretation The small peak at about 122℃ indicates the presence of a PP copolymer with ethylene. Oxidative decomposition begins well above the PP melting peak. This means the oxygen present in the air does not influence the melting curve. Nevertheless some standards require measurement of the melting peak under nitrogen.

Conclusion If an automatic gas controller is not available, the melting peak can also be measured under air. The onset of the oxidation reaction is a measure of the stability (the higher the temperature, the better the stability). Application Example "PE, Oxidation Stability" deals with the isothermal oxidation induction time, OIT.

空气中聚丙烯重复性　PP, Repeated Cycling in Air

样品 **Sample**	Polyfill PPHT5038 球状颗粒 Polyfill PPHT5038 as pellet	
条件 **Conditions**	测试仪器：DSC 坩埚： 40μl 标准铝坩埚,钻孔盖 样品制备： 从球状颗粒中心部分切出圆片,7.3 mg DSC 测试： 以 10 K/min 从 50℃加热至 220℃,然后以 5K/min 从 220℃冷却至 5℃,用同一样品重复多次。 气氛：空气,50cm³/min	Measuring cell：DSC Pan： Aluminum standard 40μl, pierced lid Sample preparation： Disk cut from center part of pellet, 7.3 mg DSC measurement： Heating from 50℃ to 220℃ at 10 K/min, then cooling from 220℃ to 50℃ at 5 K/min several times with the same sample Atmosphere：Air, 50cm³/min

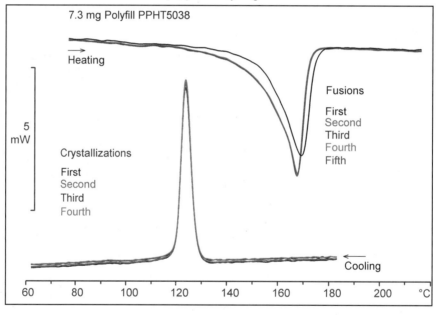

计算
Evaluation

熔融曲线 Melting curve	1	2	3	4	5	x̄	σ
结晶度,% Crystallinity in %	23.6	28.0	28.4	28.4	28.2	28.22	0.15
峰高,mW Peak height in mW	3.81	4.40	4.42	4.46	4.50	4.44	0.04
峰温,℃ Peak temperature in ℃	168.3	166.8	166.8	166.6	166.6	166.62	0.2
峰宽,K Peak width in K	11.8	11.2	11.3	11.1	11.0	11.15	0.1

结晶曲线 Crystallization curve	1	2	3	4	$\bar{x}$	σ
结晶度,% Crystallinity in %	25.6	25.3	25.3	25.8	25.29	0.26
峰高,mW Peak height in mW	6.13	6.21	6.26	6.32	6.23	0.07
峰温,℃ Peak temperature in ℃	124.4	124.4	124.4	124.4	124.40	0.0
峰宽,K Peak width in K	4.40	4.32	4.31	4.31	4.34	0.03

从四个结果（不含第一轮加热测试）计算得到的平均值 $\bar{x}$ 和标准偏差 σ。

解释 除了第一轮熔融曲线外，所有后面的扫描曲线都是一样的。对结果的计算可确定数据的重复性（标准偏差）。由于热历史不同，第一轮熔融测试没有用于统计。与结晶（冷却）曲线相比，熔融曲线的结晶度较高是因为熔融热对温度的依赖性。

结论 稳定的聚烯烃可在它们的熔程内多次加热冷却循环。由于在循环时没有发生明显的降解，所以对结果进行了统计分析。得到的标准偏差显示，测试的重复性非常好。
根据经验，我们知道不太稳定的PP在第三次循环后不会再结晶。
第一次与后面几次熔融曲线之间的不同是由于颗粒的未知热历史所造成的。

The mean values, $\bar{x}$, and the standard deviations, σ, are calculated from 4 results (without the first heating measurement).

Interpretation Except for the first melting curve, all the subsequent scans are the same. Evaluation of the results allows the reproducibility (standard deviation) of the data to be determined. The first melting measurement has not been used for the statistics because the thermal history was different. The higher degree of crystallinity of the melting curves compared with the crystallization (cooling) curves is due to the temperature dependence of the heat of fusion.

Couclusion Stabilized polyolefines can be cycled several times through their melting range. Since there is no visible degradation during cycling, statistical analysis has been performed on the results. The standard deviations obtained show the excellent reproducibility of the measurements.
From experience, we know that PP that is less stabilized would no longer crystallize after the third cycle.
The difference between the first and the following melting curves is due to the unknown thermal history of the pellet.

聚丙烯：新料与再生料 PP, New Versus Recycled Material

样品 Moplen SP 179 94 球状颗粒与据称具有相同组成的磨过的再生料（粉末）
Sample Moplen SP 179 94 as pellet and milled recycled material (powder) said to be of the same quality

条件　测试仪器：DSC　　　　　　　　　　**Measuring cell**: DSC
Conditions　坩埚：40μl 标准铝坩埚，钻孔盖　　**Pan**: Aluminum standard 40μl, pierced lid
　　　　　样品制备：　　　　　　　　　　　**Sample preparation**:
　　　　　粉末或从球状颗粒上切下的圆片　　　Powder, or disk cut from pellet
　　　　　DSC 测试：　　　　　　　　　　**DSC measurement**:
　　　　　以 10 K/min 从 30℃ 加热至 250℃　　Heating from 30℃ to 250℃ at 10 K/min
　　　　　气氛：空气，50 cm³/min　　　　　　**Atmosphere**: Air, 50 cm³/min

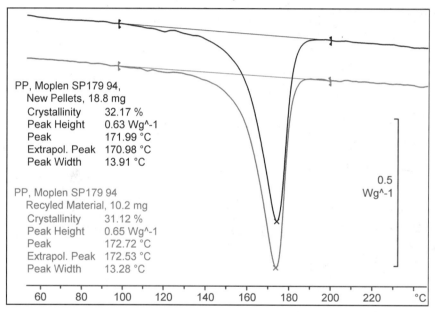

	球状颗粒 Pellets	磨过的 PP Milled PP
结晶度,% Crystallinity in %	32.2	31.1
峰温,℃ Peak temperature in ℃	172.0	172.7
峰宽,K Peak width in K	13.9	13.3

计算 / Evaluation (see table above)

解释　两条 DSC 曲线的形状一样。在聚合物数据的重复性内，计算结果是一致的。

Interpretation　The shape of the two DSC curves is identical. The evaluated results are identical within the reproducibility of polymer data.

结论　两个样品显示了相同的熔融行为。因此就 DSC 而言，两个聚丙烯样品在微晶尺寸分布、熔程、结晶度与硬度上都是相同的。

Conclusion　The two samples show the same melting behavior. As far as DSC is concerned, the two samples of polypropylene are therefore identical with respect to crystallite size distribution, melting range, crystallinity and hardness.

7.3 聚苯乙烯的玻璃化转变　Glass Transition of Polystyrene

聚苯乙烯的 DSC 曲线　DSC curves of PS

样品	Styron 683		
Sample	Styron 683		
条件	测试仪器:DSC	**Measuring cell**:DSC	
Conditions	坩埚:	**Pan**:	
	40μl 标准铝坩埚,钻孔盖	Aluminum standard 40μl, pierced lid	
	样品制备:	**Sample preparation**:	
	从球状颗粒上切出圆片,12 mg	Disk cut from pellet, 12 mg	
	DSC 测试:	**DSC measurement**:	
	第一轮加热以 10 K/min 从 30℃至 250℃,在自动进样器中骤冷至室温,第二轮加热以 10 K/min 从 30℃加热至 250℃	First heating run from 30℃ to 250℃ at 10 K/min, quench-cooled to room temperature in the sample changer, second heating run from 30℃ to 250℃ at 10 K/min.	
	气氛:氮气,50 cm^3/min	**Atmosphere**:Nitrogen, 50 cm^3/min	

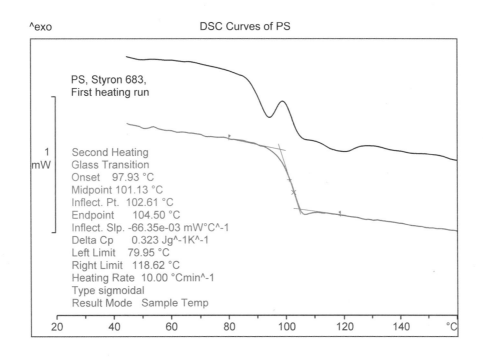

计算　第一轮没有计算,因为由于热历史效应而有变形。第二轮用玻璃化转变显示结果及其他信息。转变相当清晰(起始点与终点只有 7K,范围很小)。比热的变化为 0.32 J/gK。

Evaluation　The first run has not been evaluated because it is distorted due to the effects of thermal history. The second run shows the results and additonal information given by the glass transition evaluation. The transition is quite sharp (indicated by the small range of 7 K between onset and end point). The change of the specific heat capacity is 0.32 J/gK.

解释 聚苯乙烯，至少市售的无规型，是完全无定形的。它的玻璃化转变温度约为100℃。通常第一轮加热曲线由于松驰过程而在玻璃化转变区是不规则的。第二轮加热由于没有热历史，生成正常的玻璃化转变曲线。

结论 应该对第二轮加热进行数据计算。

Interpretation Polystyrene, at least the commercially available atactic type, is entirely amorphous. Its glass transition temperature is approx. 100℃. Usually the first heating run gives a curve in which the region around the glass transition is distorted due to relaxation processes. The second heating run without the thermal history yields the normal shape of a glass transition.

Conclusion The evaluation should be performed on the second heating run.

用 DLTMA 测定聚苯乙烯的玻璃化转变   PS, Glass transition by DLTMA

样品　聚苯乙烯,透明薄片,厚 0.87 mm
Sample　Polystyrene, transparent sheet 0.87 mm thick

条件　测试仪器:
Conditions　TMA,DLTMA（动态负载 TMA）模式,3mm 圆点探头

Measuring cell: TMA in the mode DLTMA (dynamic load TMA) with 3-mm ball-point probe

坩埚:
无需坩埚；样品直接放在石英样品架上

Pan: No pan; the specimen was placed directly on the fused silica sample support

样品制备:
用钳子从脆性片上钳下一片；质量约为 12 mg

Sample preparation: A piece was broken off the brittle sheet using pliers; mass approx. 12 mg

测试:
第一轮加热以 10 K/min 从 25℃ 至 145℃,在动态负载下自由冷却至室温,第二轮加热以 5 K/min 从 25℃ 至 135℃

Measurement: First heating run from 25℃ to 145℃ at 10 K/min, uncontrolled cooling to room temperature under dynamic load, second heating run from 25℃ to 135℃ at 5 K/min

负载:
方形波,负载每 6 秒从 0.05 N 变化至 0.25 N

Load: Square wave with the load changing from 0.05 N to 0.25 N every 6 s

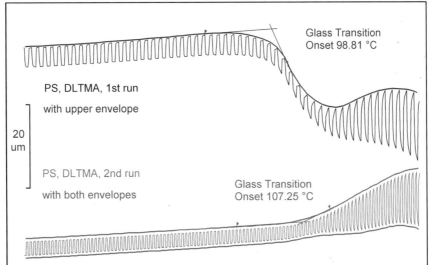

数据处理　第一轮加热曲线的穿透（凹陷）起始点为 98.8 ℃。
通常,推荐第二轮进行计算。它的上包络线给出：

Evaluation　The onset of penetration (indentation) in the first heating curve is 98.8 ℃.
Often, the second run is preferred for evaluations. Its upper envelope gives:

69

- T_g as the onset in the region of the transition of 107.3 ℃.
- Expansivities based on the slope: below T_g: 58 ppm/K, above T_g: 530 ppm/K.

The difference of the envelopes is proportional to the compliance (reciprocal Young's modulus). In the diagram, for sake of clarity, the envelopes are slightly sifted with respect to the curve.

Intrepretation The first run shows the change in thickness caused by the change in the load. The sample is compressed by the load of 0.25 N. It recovers during the following 6 s when the load is 0.05 N ($2\mu m$ displacement is caused by the spring constant of the TMA of approx. $10\mu m/N$).

— The slope of the envelope up to 80 ℃ corresponds to the expansivity below the glass transition temperature. The sample is rigid enough to carry the 3 mm sphere under the mean load of 0.15 N. In reality, the sphere does not make contact with the sample at a single point, but makes an indentation with a larger area of contact.

— The compressive stress under a freshly placed sphere always corresponds to the ball thrust hardness of the specimen (PS≈110 N/mm²). With a load of 0.15 N, the sphere indents as much as required to give a contact area of 1.4 10⁻³ mm² corresponding to a diameter of $40\mu m$.

— Between 80 ℃ and 120 ℃ the probe falls a further $20\mu m$. This effect commonly observed on the first heating curve has different causes:

- The probe indents into the softening sample until the increasing contact area is again able to support the sphere (the ball thrust hardness decreases by orders of magnitude around T_g).
- Minor unevenness of the bottom surface of the sample adapts to the sample support on softening.

Sometimes an inverse effect, a kind of swelling is visible that is caused by frozen stress or volume relaxation (corresponding to the DSC enthalpy relaxation).

Above 110 ℃, the increasing strain amplitude indicates the decreasing Young's modulus. At 135℃, plastic deformation begins and causes the indentation of the ball-point probe at 150℃ to reach a ΔL of $40\mu m$ corresponding to a calotte of 0.4 mm² ($2\pi r\Delta L$) or a diameter of the contact area of approx. 0.7

The second run begins with completely changed geometry: the diameter of the contact area increased from 0.04 to 0.7 mm. The thermal history is also different because the sample has been cooled under dynamic load.

The mean compressive stress under the sphere is only 0.4 N/mm². That is why the second curve continuously increases (at > 150 ℃ there would be plastic deformation again until the probe has penetrated the entire sample).

A close look at the DLTMA curve shows the fast sample response (square wave) below the T_g. Above T_g the behavior is viscoelastic and the response therefore slower. In this case, the 6 s are not long enough for equilibrium deformation to be reached.

Conclusion Although the physics behind these events is relatively complex, the experiments are easy to perform. If the elastic response is not required, a normal TMA experiment at constant load is sufficient.

Sometimes, the glass transition is not clearly visible in a DSC curve. In such cases TMA is advantageous. Even with highly filled polymers, where DSC may give transitions that are questionable or not visible, DLTMA furnishes clear results, at least in the three-point bending mode.

7.4 聚氯乙烯的热分析测试 TA Measurements on Polyvinyl Chloride

用 DSC 和 TGA 测试聚氯乙烯 PVC Measured by DSC and TGA

样品	未增塑 PVC Trosiplast 3255 的注射成型板材
Sample	Unplasticized PVC, Trosiplast 3255 injection-molded sheet material

$$\left[\begin{array}{c} Cl \\ | \\ -CH_2-CH_2- \end{array} \right]_n$$

条件 / Conditions

测试仪器：DSC，TGA
Measuring cell: DSC, TGA

坩埚：
DSC：40 μl 标准铝坩埚，钻孔盖
TGA：70 μl 氧化铝坩埚，无盖

Pan:
DSC: Aluminum standard 40 μl, pierced lid
TGA: Alumina 70 μl without lid

样品制备：
从板材上切下小片

Sample preparation:
Disk cut off sheet

DSC 测试：
以 10 K/min 从 30℃加热至 120℃，在自动进样器上骤冷第二轮以 10 K/min 从 30℃加热至 300℃

DSC measurement:
Heating from 30℃ to 120℃ at 10 K/min, quench-cooled in the sample changer Second run from 30℃ to 300 ℃ at 10 K/min

TGA 测试：
以 10 K/min 从 30℃加热至 300℃
气氛：氮气，50 cm³/min

TGA measurement:
Heating from 30℃ to 300℃ at 10 K/min
Atmosphere: Nitrogen, 50 cm³/min

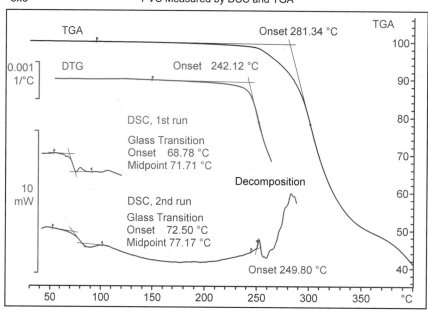

计算
Evaluation

玻璃化温度,起始点,第一轮,℃ Glass temperature, onset, first run in ℃	68.8
玻璃化温度,起始点,第二轮,℃ Glass temperature, onset, second run in ℃	72.5
DSC 分解起始点,℃ Decomposition onset DSC in ℃	250.0
TGA 分解起始点,℃ Decomposition onset TGA in ℃	281.0
DTG 分解起始点,℃ Decomposition onset DTG in ℃	242.0

所有的起始点都基于拐切线。　　　　　　All onsets are based on the inflectional tangents.

解释　第一轮 DSC 曲线显示典型的不连续性,它们是由于应力松弛和焓松弛造成的。通常第二轮得到没有任何干扰的玻璃化转变。

在大约 200℃ 以上,热降解通常开始,DSC 上为放热峰,TGA 测到的始终为由于 HCl 生成导致的失重。HCl 会侵蚀测试池。用氮气吹扫测试池是防止损坏的防范措施。如果只研究分解的开始,则可以中断温度程序,将仪器冷却下来。为了比较,第一个分解峰可作为起始点。放热分解部分是由 HCl 与铝坩埚的化学反应造成的。

结论　PVC 是对热敏感的聚合物。它分解生成 HCl。
典型特征是玻璃化转变温度和分解温度(它是对热稳定性的量度)。

Interpretation　The first DSC curve shows typical discontinuities. They are caused by stress re-lief and enthalpy relaxation. Usually the second run gives glass transitions free of any disturbances.

Above about 200℃, thermal degradation often begins in the DSC with an exo-thermal peak and always with weight loss due to the formation of HCl, measured by TGA. The HCl could attack the measuring cell. Purging the cell with nitrogen is a countermeasure to prevent damage. If only the beginning of decomposition is to be studied, the temperature program can be interrupted and the cell cooled down. The first decomposition peak is evaluated as onset for comparison pur-poses.

Part of the exothermal decomposition is caused by the chemical reaction of the HCl with the aluminum pan.

Conclusion　PVC is a polymer that is sensitive to heat. It decomposes with the formation of HCl.

The characteristic features are the glass transition temperature and the temperature at which degradation begins (which is a measure of its stability toward heat).

未增塑聚氯乙烯的热稳定性　PVC-U, Thermal stability

样品 由德国塑料工业加工协会提供的一套未增塑 PVC 薄膜样品
Sample Unplasticized PVC film supplied in the set of samples of the German "Arbeitsgemeinschaft Deutsche Kunststoff-Industrie"

条件 测试仪器：TGA
Conditions

坩埚：
70 μl 铝坩埚，无盖

样品制备：
用剪子剪成片：8.937 mg (5 K/min)，8.558 mg (10 K/min)，8.974 mg (15 K/min) 和 8.935 mg (20 K/min)

TGA 测试：
分别以 5、10、15 和 20 K/min 从 25℃加热，最终温度 500℃。

气氛：氮气，50 cm³/min

Measuring cell：TGA

Pan：
Alumina 70 μl, without lid

Sample preparation：
Disks cut with scissors：8.937 mg (5 K/min)，8.558 mg (10 K/min)，8.974 mg (15 K/min) and 8.935 mg (20 K/min)

TGA measurement：
Heating from 25℃ at 5, 10, 15 and 20 K/min respectively. Final temperature 500℃

Atmosphere：Nitrogen，50 cm³/min

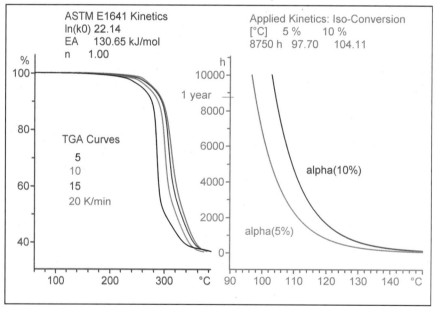

计算　数据处理按照 ASTM E 1641 计算对应某一转化率如 10% 的温度。由于在计算热稳定性中转化率函数会产生误差，所以应考虑 20% 为最大值。用非模型动力学进行数据处理时，对大于 20% 的分解程度可以给出很好的预测。本例以 5% 的转化率（即质量减少 3.15%）为基础。

Evaluation　The evaluation according to ASTM E 1641 calculates temperatures corresponding to certain conversions, e.g. 10%. A maximum value of 20% should be considered since the function of conversion would cause errors in the calculated thermal stability. Evaluating with Model Free Kinetics would yield good predictions for extents of decomposition >20%. This example is based on 5% decomposition (i.e. a 3.15% decrease in mass).

The kinetic results are:

$\ln k_0$ 22.14 E_a 130.65 kJ/mol

According to ASTM E 1641, predictions for the thermal endurance are made using a first order kinetic model because the error of calculation is small below 20% decomposition. Normally the predictions are made for the conversion that has been used in the calculation of the kinetic data. As the diagram shows, reaction times (time to failure) can be calculated for other conversions, α, e.g. of 10%.

The two curves plotted for identical conversion (isoconversion curves) for 5% and 10% decomposition show the thermal endurance in the range of 500 to 10000 hours (more than one year). The table gives the temperatures corresponding to a time to failure of one year for 5% and 10% decomposition (97.7 ℃ and 104.1 ℃ respectively).

Interpretation The TGA curves (on the left) show the weight loss due to the elimination of HCl. As expected, the curves are shifted to higher temperatures with increasing heating rate (kinetic effect). The weight loss of 63% is greater than the stoichiometrically expected value for HCl of 58.4%. There are some other volatile components.

Conclusion The thermal endurance is estimated on the basis of four TGA curves measured at heating different rates. ASTM E 1641 assumes a rate of decomposition that decreases with decreasing temperature according to Arrhenius. There is a risk of errors arising through extrapolation. A countermeasure is the use of very low heating rates with correspondingly low decomposition temperatures. To save time, only the measurement with the highest rate is run to complete decomposition. All others are stopped as soon as a threshold value of 5% is reached.

聚氯乙烯的 TMA 曲线与所加负载的关系
PVC, TMA curves as a function of applied load

样品 **Sample**	聚氯乙烯圆片，直径约为 3 mm Disk of polyvinyl chloride of approx. 3 mm diameter	
条件 **Conditions**	测试仪器： TMA，3 mm 圆点探头	Measuring cell： TMA with 3 mm ball-point probe
	坩埚： 样品放置在直径 6mm、厚 0.5mm 的石英圆片上	Pan： The sample is placed on a fused silica disk of 6 mm diameter and 0.5 mm thickness.
	样品制备： 每次测试都用一个新样品	Sample preparation： Each measurement was performed with a new sample
	TMA 测试： 以 5 K/min 从 30℃加热到 130℃。	TMA measurement： Heating from 30℃ to 130℃ at 5 K/min
	负载： 0.01N、0.1N、0.5N、1N 和 2N。每次测试都使用一片新的 PVC 圆片。使用附加重量得到大于0.5N的负载。	Load： 0.01 N, 0.1 N, 0.5 N, 1 N and 2 N. For each measurement, a new PVC disk was used. Loads >0.5 N were obtained by using additional weights
	气氛：静态空气	Atmosphere：Static air

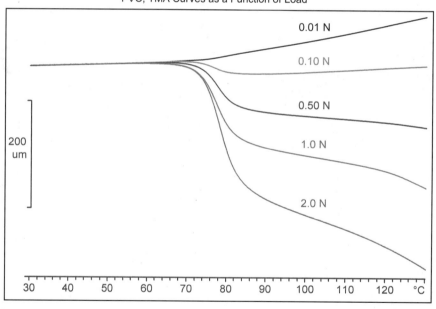

计算 切线的交点始于 60℃和 76℃，40℃与 50℃之间的平均膨胀率 α_{mean} 归纳在下表中。为了清晰起见，图中没有显示单个的数据处理。

Evaluation The intersection points of the tangents drawn at 60℃ and 76℃ and the mean expansivity, α_{mean}, between 40℃ and 50℃ are summarized in the table. For the sake of clarity, the individual evaluations are not shown in the figure.

负载,N Load in N	0.01	0.1	0.5	1.0	2.0
起始点,℃ Onset in ℃	72.8	72.3	72.6	72.2	72.7
α_{mean},ppm/K	76.4	—	—	—	—

只有当使用较小的外力时,膨胀率的测试才能基本上不受压缩影响。力较大时,在样品与圆点探头间放置一个石英圆片,将压缩应力均匀地分散到大的截面上。

解释 最小负载的曲线实质上是一个膨胀测试。圆球产生的压缩应力等于球抗压硬度(25℃下PVC约为70 N/mm²)。凹陷和相应的接触面积足以支撑圆球。负载越大,凹陷越大。玻璃化转变温度处硬度的显著下降将产生更大的凹陷。

注意:只有第一轮曲线给出软化时的凹陷。如果样品在各自相应的负载下冷却并再加热,冻结的应力会支撑住圆球,不会产生任何进一步的凹陷。

结论 用圆点探头进行的TMA测试给出实际上与所加负载无关的可重复的软化温度。在较高负载下,如0.5 N,由于基线与切线间的角度较大,可更好地确定交点。

The expansivity can only be measured essentially free of the effects of compression when low forces are used. With higher forces, the compressive stress would have to be evenly distributed over a large cross-sectional area by placing a fused silica disk between sample and ball-point probe.

Interpretation The curve with the smallest load is essentially a dilatometric measurement. The compressive stress caused by a ball is equal to the ball thrust hardness (approx. 70 N/mm² for PVC at 25℃). The indentation and the corresponding contact area is just sufficient to support the ball. The higher the load, the higher the indentation. The dramatic decrease of the stiffness at the glass transition temperature causes the additional indentation.

Note: Only the first curve gives the indentation on softening. If the sample is cooled under the respective load and reheated, the frozen stress can carry the ball without any further indentation.

Conclusion TMA measurements with a ball-point probe give reproducible softening temperatures that are practically independent of the applied load. With higher loads of, e.g. 0.5 N, the intersection point is better defined due to the greater angle between baseline and tangent.

聚氯乙烯和氯化聚氯乙烯的玻璃化转变 Glass transition of PVC and chlorinated PVC

样品
1. 含有 56.6% Cl 的 PVC 粉末, 8.792 mg
2. 含有 60.8% Cl 的氯化 PVC(PVCC), 14.903 mg
3. 含有 66.4% Cl 的氯化 PVC(PVCC), 10.804 mg

Sample
1. PVC powder with 56.6% Cl, 8.792 mg
2. Chlorinated PVC (PVCC) with 60.8% Cl, 14.903 mg
3. Chlorinated PVC (PVCC) with 66.4% Cl, 10.804 mg

条件 / Conditions

测试仪器: DSC
Measuring cell: DSC

坩埚:
40 μl 标准铝坩埚,密封
Pan:
Aluminum standard 40 μl, hermetically sealed

样品制备:
在密封前不压称量样品
Sample preparation:
Samples weighed in without pressing before sealing

DSC 测试:
以 10 K/min 从 35℃ 加热到 150℃ 以消除热历史,在自动进样器上骤冷
第二轮以 10 K/min 从 35℃ 加热到 150℃

DSC measurement:
Heating from 35℃ to 150℃ at 10 K/min to eliminate thermal history, quench-cooling in the sample changer
Second heating run from 35℃ to 150℃ at 10 K/min

气氛: 静态空气
Atmosphere: Static air

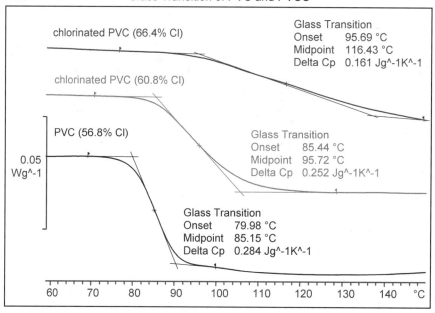

计算 / Evaluation

样品号 Sample No	质量, mg Mass in mg	氯含量, % Chlorine content in %	T_g, 中点, ℃ T_g, mid point in ℃	Δc_p, J/gK
1	8.792	56.6	85.2	0.28
2	14.903	60.8	95.7	0.25
3	10.804	66.4	116.4	0.16

Interpretation The three measurements show the effect of the chlorine concentration on the glass transition. Higher concentrations of chlorine decrease the molecular mobility. As a result of this, the glass transition shifts to higher temperatures. The broadening of the glass transition with increasing chlorine content is particularly noticeable. The reason for this is the relatively large degree of inhomogeneity of the chlorine distribution. In chlorination, a hydrogen atom is replaced by a chlorine atom. This does not change the number of degrees of freedom of a monomer unit. The step height (Δc_p) with respect to the molar mass therefore remains unaffected by chlorination. The decrease in the step height with increasing chlorination, which is apparent in the figure, is therefore due to the increase in size of the molar mass.

This allows the change of Δc_p to be used to estimate the chlorine content. The molar mass of a PVC monomer unit, M_{PVC}, is 62.5 g/mol. Because the molar mass of chlorine is 35.5 g/mol, this gives a value of 56.8% for the chlorine content of PVC. The Δc_p step height, Δc_{pPVC} is 0.28 J/gK. This corresponds to 17.5 J/molK. The height of the Δc_p step of the chlorinated PVC sample with the lower content of chlorine can determined relatively accurately (Δc_{pPVCC} 0.24 J/gK). The molar mass of the chlorinated PVC, M_{PVCC}, can be estimated from the equation

$$M_{PVCC} = M_{PVC}(\Delta C_{pPVC} / \Delta C_{pPVCC})$$

In the case considered, this gives a value of M_{PVCC} of 70.0 g/mol. This corresponds to 1.21 chlorine atoms per monomer unit and hence to a chlorine content of 61.4%. This agrees very well with the data given for this sample.

Conclusion The results show the influence of increasing chlorine content: the glass transition temperature increases, the range broadens and the change in specific heat capacity becomes smaller. These results are useful for quality control purposes.

聚氯乙烯增塑剂混合物的凝胶化　Gelation of a PVC Plasticizer Mixture

样品　　PVC 粉末增塑剂混合物((Fimo® 成模材料,Faber GmbH)
　　　　　增塑的 PVC 是由 PVC 粉末增塑剂混合物的热凝胶制成。本实验的目的是测定凝胶化的温度范围与最终产物的玻璃化转变范围。

Sample　PVC powder plasticizer mixture (Fimo® modeling material, Faber GmbH).
Plasticized PVC is made by thermal gelation of a PVC powder plasticizer mixture. The purpose of the experiment was to determine the temperature range of gelation as well as the glass transition range of the final product.

条件　　测试仪器:
Conditions　DMA,剪切夹具支架

Measuring cell:
DMA with the shear clamp holder

样品制备:
做成两个重约 130 mg 的球,用剪切夹具的固定螺钉挤压成直径约 10mm 的 1mm 厚圆片。

Sample preparation:
Two balls of approx. 130 mg are formed and squeezed to 1 mm disks of approx. 10 mm diameter using the fixation screws of the shear clamps.

DMA 测试:
第一轮加热以 3 K/min 从 25℃ 到 140℃,然后对同一样品重新夹紧后从 −25℃ 开始进行第二轮加热测试。
使用多频模式,对样品以 1、2、5 和 10 Hz 的频率进行同步测试。最大力振幅 1 N;最大位移振幅 10 μm;偏移控制为零。

DMA measurement:
First heating run from 25℃ to 140℃ at 3 K/min followed by a second heating measurement of the same specimen (the resulting plasticized PVC) after reclamping and starting at −25℃.
The multi-frequency mode is used in which the sample is simultaneously measured at frequencies of 1, 2, 5 and 10 Hz. Maximum force amplitude 1 N; maximum displacement amplitude 10 μm; offset control zero.

气氛:静态空气　　　Atmosphere: Static Air

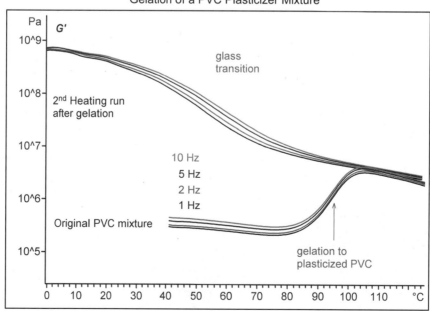

Gelation of a PVC Plasticizer Mixture

解释 剪切模量约为 20 MPa 的初始的粘流态混合物在 80℃至 105℃之间经历凝胶化。在这个过程中,模量增加超过一个数量级。不同频率下测试的曲线显示该过程不是频率依赖性的,与玻璃化转变是不同的。在冷却到玻璃态时,生成的增塑物的模量达到了 500 MPa。材料此时已足够坚硬并能保持形状。第二轮加热显示在 30℃至 70℃间的宽玻璃化转变具有典型的频率依赖性。不同增塑剂混合物是造成很宽玻璃化转变的原因。作为比较,未塑化 PVC 的玻璃化转变温度为 80℃。

结论 配剪切夹具支架的 DMA 对从测试粘流态至玻璃态变化的样品是理想的。使用不同的频率有助于解释用 DMA 测得的结果。

Interpretation The initially viscous mixture with a shear modulus of approx. 20 MPa undergoes gelation between 80℃ and 105℃. During this process the modulus increases by more than a decade. The curves measured at different frequencies show that the process is not frequency dependent and that it is different to a glass transition. On cooling to the glassy state, the modulus of the resulting plasticized reaches 500 MPa. The material is now rigid enough to maintain a given shape. The second heating run shows the broad glass transition between 30℃ and 70℃ with typical frequency dependence. The mixture of different plasticizers is the reason for the very broad glass transition. In comparison, the glass transition of unplasticized PVC occurs at 80℃.

Conclusion The DMA with the shear clamp holder is ideal for measurements in which the sample changes from the viscous to the glassy state. The use of different frequencies helps to interpret the effects measured by DMA.

7.5 聚酰胺及其共混物 Polyamides and their blends

聚酰胺 6 的熔融行为 Polyamide 6, Melting Behavior

样品	Ultramid B3G5HSQ15 球状颗粒和再生料	$\left[-NH-(CH_2)_5-\overset{\overset{O}{\|\|}}{C}-NH-(CH_2)_5-\overset{\overset{O}{\|\|}}{C}-\right]_n$
Sample	Ultramid B3G5HSQ15 as pellets and recycled material	
条件 **Conditions**	测试仪器:DSC	**Measuring cell**:DSC
	坩埚:40 μl 标准铝坩埚,钻孔盖	**Pan**:Aluminum standard 40 μl, pierced lid
	样品制备:	**Sample preparation**:
	从球状颗粒中心部分切出圆片(7.5 mg);9 mg 粉末	Disk cut from center part pellet (7.5 mg), 9 mg powder
	DSC 测试:	**DSC measurement**:
	以 20 K/min 从 30℃加热到 300℃, 以 20 K/min 冷却到 100℃	Heating from 30℃ to 300℃ at 20 K/min, cooling to 100℃ at 20 K/min
	第二轮加热以 20 K/min 从 30℃到 300℃	Second heating run from 30℃ to 300℃ at 20 K/min
	气氛:氮气,50 cm³/min	**Atmosphere**:Nitrogen,50 cm³/min

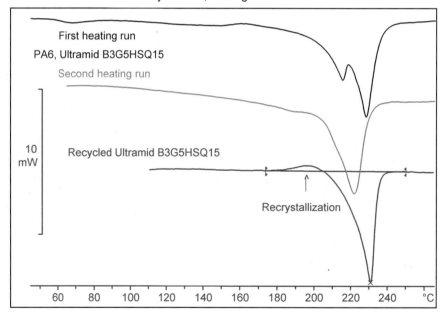

计算 **Evaluation**	熔融峰的积分结果如下: The results of the integration of the melting peaks are given below:

	颗粒 Pellet	颗粒 Pellet	再生料 Recycled material
	第一轮加热 1st heating run	第二轮加热 2nd heating run	第一轮加热 1st heating run
积分,J/g Integral in J/g	55.7	49.6	55.5
峰温,℃ Peak temperature in ℃	227.2	220.8	229.7

注：如果发生再结晶，积分时必须包含该放热峰。得到的熔融热对应于没有再结晶的原始结晶度，因为微晶形成所释放的能量等于其熔融的能量。为了测定可能的最大结晶度，样品应该在放热预熔融峰出现后立即冷却约 50 K。然后用后面完整的加热速率曲线来测量结晶度。

解释 第一轮熔融曲线经常显示出双峰。可能的原因是：
- 不规则的热传递（取决于样品的形状）
- 多晶态（六角形微晶在 215℃ 熔融，单斜晶部分在 220℃ 熔融）
- 微晶分离（热历史，见应用实例"聚乙烯的熔融曲线与热历史"）

50℃ 到 90℃ 范围内的小吸热效应是由于曾暴露于大气湿度下的聚酰胺颗粒的干燥产生。

再生料熔融峰前的放热峰是由于所谓的再结晶：如果分子的活动性足够高（活动性随着温度升高而增加），没有达到它最大结晶度的聚合物会结晶。

结论 第一轮DSC熔融峰受到样品热历史的影响。为了表征材料，样品在第一轮测试后应该在设定的条件下冷却，然后作第二轮测试。

Note: If recrystallization occurs, the integral must include this exothermic peak. The resulting heat of fusion corresponds to the original crystallinity without recrystallization because the energy released by the formation of the crystallites is equal to the energy required to melt them. To determine the maximum possible degree of crystallinity, the sample should be cooled by approx. 50 K immediately after the exothermic premelt peak occurs. The subsequent complete heating curve can then used to measure the crystallinity.

Interpretation The first melting curve often shows a double peak. Possible causes are
- irregular heat transfer (depending on the shape of the sample),
- polymorphism (hexagonal crystallites melt at 215℃, the monoclinic fraction at 220℃)
- crystallite segregation (thermal history, see Application Example "PE, melting curve and thermal history").

The small endothermic event in the range 50℃ to 90℃ is due to the drying of polyamide granules that had been exposed to atmospheric humidity.

The exothermic peak of the recycled material prior to melting is due to so-called recrystallization: a polymer that has not yet reached its maximum crystallinity forms crystallites if the mobility of the molecules is sufficiently high (mobility increases with temperature).

Conclusions The first DSC melting peak is influenced by the thermal history of the sample. To characterize the material, the sample should be cooled under defined conditions after the first measurement and then measured a second time.

聚酰胺 6：新料与再生料　　PA6, New Versus Recycled Material

样品 **Sample**	新的 Durethan BKV 15 球状颗粒与 Durethan BKV 15 再生料 Durethan BKV 15, new pellets and Durethan BKV 15 recycled material	
条件 **Conditions**	测试仪器：DSC	**Measuring cell**：DSC
	坩埚：40 μl 标准铝坩埚，钻孔盖	**Pan**：Aluminum standard 40 μl, pierced lid
	样品制备：	Sample preparation：
	从球状颗粒中心部分切出圆片	Disk cut from center part of pellet
	DSC 测试：	DSC measurement：
	以 10 K/min 从 100℃ 加热至 300℃，以 10 K/min 冷却至 100℃（以消除热历史）	Heating from 100℃ to 300℃ at 10 K/min, cooling to 100℃ at 10 K/min (to eliminate thermal history)
	实际测试：	Actual measurement：
	以 10 K/min 从 100℃加热到 300℃	Heating from 100℃ to 300℃ at 10 K/min
	气氛：氮气，50 cm³/min	Atmosphere：Nitrogen, 50 cm³/min

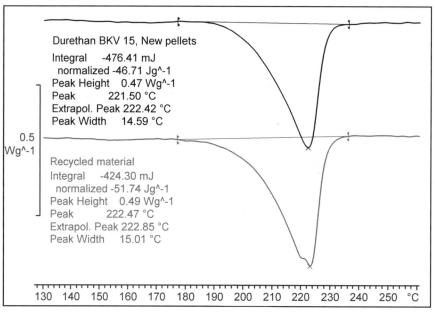

计算 熔融峰积分结果：
Evaluation　The results of the integrated melting peaks are：

样品 Sample	颗粒 Pellet	再生料 Recycled material
熔融热，J/g Heat of fusion in J/g	46.7	51.7
峰温，℃ Peak temperature in ℃	221.5	222.5
峰宽，K Peak width in K	14.6	15.0
样品质量，mg Sample mass in mg	10.2	8.2

解释 由第二轮加热的 DSC 曲线实际上是一样的。这表明两个都是 PA6。

结论 再生料的熔融热较高。再生料的结晶度实际上常高于新料，这是由于作为成核点的少量杂质的存在所造成的。由于结晶度影响机械性能，建议公司添加再生料不要超过 10%，并缩短注射成型的冷却周期。

Interpretation The DSC curves from the second heating runs are practically identical. This shows that both are PA6.

Conclusions The heat of fusion of the recycled material is higher. The degree of crystallinity of recycled material is in fact often higher than that of new material. This is due to the presence of small amounts of impurities that act as nucleation points. Since the degree of crystallinity influences mechanical properties, the company was advised not to add more than 10% of recycled material and to shorten the injection molding cooling cycle.

玻璃纤维含量的测定 Determination of Glass Fiber Content

样品 **Sample**	含和不含玻璃纤维的聚酰胺 6 球状颗粒 Polyamide 6 pellets with and without glass fibers		
条件 **Conditions**	测试仪器：TGA 坩埚： 氧化铝，70μl，无盖 样品制备： 从球状颗粒上切下的圆片 TGA 测试： 以 10 K/ming 从 25℃加热到 700℃ 气氛：氮气，50 cm³/min	**Measuring cell**：TGA **Pan**： Alumina，70μl，no lid **Sample preparation**： Disk cut from pellet **TGA measurement**： Heating from 25℃ to 700℃ at 10 K/min **Atmosphere**：Nitrogen，50 cm³/min	

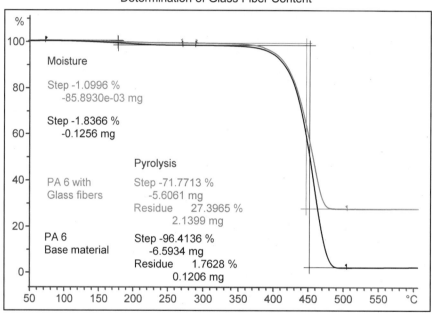

计算 500℃以上，残余物 R 如下：
PA6，基材，　　R_{PA6}：1.76%
PA6，纤维增强，R_{GFRP}：27.40%
从两种残余物的差值可估算玻璃纤维含量 α_{GF}（非直接分析）：

Evaluation Above 500 ℃, the residues, R, are as follows：
PA6, base material, 　　R_{PA6}：1.76%
PA6, fiber-reinforced, R_{GFRP}：27.40%.
The glass fiber content α_{GF} can be estimated from the difference of the two residues (indirect analysis)：

$$\alpha_{GF} = [(R_{GFRP} - R_{PA6})/(100 - R_{PA6})] \cdot 100\% = 26.1\%$$

解释 两个样品都显示出两个失重台阶。第一个是由失水产生的——样品曾置于环境空气中。玻璃纤维增强的样品吸收较少的水分,因为它的聚合物含量较低。在热解台阶后,两个样品都留下残余物。在玻璃纤维增强样品中,该残余主要由玻璃纤维组成。

结论 用TGA进行成分分析能测定PA6样品中的水份、聚合物和玻璃纤维的含量。TGA聚合物热解曲线上的拐点能用作鉴定。

Interpretation Both samples exhibit two weight loss steps. The first is caused by the elimination of moisture-the samples had been kept in ambient air. The glass-fiber reinforced sample had absorbed less water because its polymer content was lower. After the pyrolysis step, both samples left a residue. In the case of the glass-fiber reinforced sample, this residue consisted mainly of glass fibers.

Conclusions Compositional analysis by TGA can determine the moisture, polymer and glass fiber contents in a PA6 sample. The inflection point in the TGA polymer pyrolysis curve can be used for identification purposes.

不同质量的聚酰胺66 PA66, Different Qualities

样品　　Ultramid C3ZM6（球状颗粒），
　　　　　　Bergamid A70B30（成型部件）和
　　　　　　Ultramid A3K（成型部件）

Sample　Ultramid C3ZM6 (pellets),
　　　　　　Bergamid A70B30 (molded part) and Ultramid A3K (molded part)

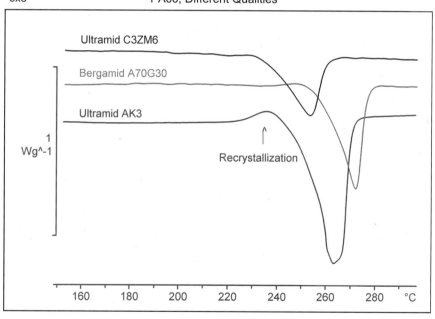

条件	测试仪器：DSC
Conditions	**Measuring cell**：DSC
	坩埚：
	Pan：
	40μl 标准铝坩埚，钻孔盖
	Aluminum standard 40μl, pierced lid
	样品制备：
	Sample preparation：
	从球状颗粒/成型部件的中心部分切出圆片
	Disk cut from center part of pellet/molded part
	DSC 测试：
	DSC measurement：
	以 10 K/min 从 50℃加热到 300℃
	Heating from 50℃ to 300℃ at 10 K/min
	气氛：氮气，50 cm^3/min
	Atmosphere：Nitrogen，50 cm^3/min

数据处理　结果总结如下：

Evaluation　The results are summarized below：

样品 Sample	Ultramid C3ZM6	Bergamid A70B30	Ultramid A3K
积分，J/g Integral in J/g	29.7	40.7	68.9
峰温，℃ Peak temperature in ℃	252.7	271.6	262.5
样品质量，mg Sample mass in mg	12.0	6.2	11.3

注：如果再结晶发生，积分基线必须置于该放热峰前。这样该熔融热对应于没有再结晶即原始料的结晶度，因为相同量的材料的结晶热与熔融热相等。

解释 DSC 的峰温取决于微晶的大小。后者主要受控于聚合物分子的规整性。不过，结晶条件与聚合度也有影响。例如，对于 PE，峰温变动范围从约 110℃ 至 150℃。

Bergamid 是一种高分子量的 PA66，形成较大的微晶，峰温高。Ultramid C3ZM6 是含有极少 PA6 的共聚物，观察到它的熔融峰在 253℃。

结论 即使是在 PA66 族中，也可以通过产品的熔程、预熔融峰和熔融热来区分材料的不同类型。

Note: If recrystallization occurs, the baseline for integration must be set before this exothermic peak. The heat of fusion then corresponds to the crystallinity without recrystallization, i. e. to the material as delivered, since the heat of crystallization is equal to the heat of fusion of the same amount of material.

Interpretation The DSC peak temperature depends on the size of the crystallites. The latter is controlled primarily by the regularity of polymer molecule. The conditions of crystallization and the degree of polymerization, however, also have an effect. For example, for PE, the peak temperature ranges from about 110℃ to 150℃.

Bergamid is a high molecular weight PA66 that forms large crystallites with a high peak temperature. Ultramid C3ZM6 is a copolymer containing very little PA6. Its melting peak at 253 ℃ can just be seen.

Conclusion Even within the PA66 family, different types of material can be distinguished through their melting range, premelting peaks and the heat of fusion.

用 TGA 和 DSC 测定聚酰胺 66 的水含量
Determination of the Moisture Content of PA66 by TGA and DSC

样品	Ultramid C3ZM6（球状颗粒），Bergamid A70B30（成型部件）和 Ultramid A3K（成型部件）	
Sample	Ultramid C3ZM6 (pellet), Bergamid A70B30 (molded part) and Ultramid A3K (molded part)	

条件 / Conditions

测试仪器：DSC TGA	**Measuring cell**：DSC TGA
坩埚：	**Pan**：
40 μl 标准铝坩埚，钻孔盖	Aluminum standard 40 μl, pierced lid
TGA 用 70 μl 氧化铝坩埚	Alumina 70 μl for TGA
样品制备：	**Sample preparation**：
切成圆片，测试前在水中浸几个小时	Disks were cut and soaked for several hours in water prior to measurement
DSC 测试：	**DSC measurement**：
以 10 K/min 从 30℃加热到 300℃，在自动进样器上骤冷到室温，接着第二轮加热以 10 K/min 从 30℃到 300℃	Heating from 30℃ to 300℃ at 10 K/min, quench-cooling to room temperature in the sample changer, followed by a second heating run from 30℃ to 300℃ at 10 K/min
TGA 测试：	**TGA measurement**：
以 10 K/min 从 30℃加热到 400℃	Heating from 30℃ to 400℃ at 10 K/min
气氛：氮气，50 cm³/min	**Atmosphere**：Nitrogen, 50 cm³/min

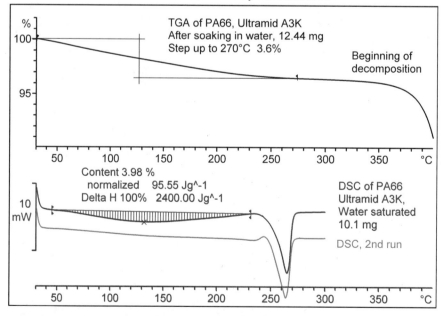

计算 Evaluation	PA66,湿 PA66, moist	PA66,干 PA66, dry
基于 TGA 台阶的水分,% Moisture based on TGA step in %	3.6	—
从 DSC 峰计算的水分,% Moisture calculated from DSC peak in %	4.0	—
峰温,℃ Peak temperature in ℃	264.0	262.8

水的蒸发热 2400 J/g 来计算 DSC 的水分含量。

A heat of evaporation of water of 2400 J/g was used to calculate the moisture content by DSC.

解释 测试显示,所吸收的水分较缓慢地从 1 mm 厚的样品上蒸发。TGA 失重台阶和 DSC 吸热峰相应的比较宽。300℃以上的失重增加是由聚酰胺降解开始产生的。

第二轮 DSC 曲线轨迹上可见的小吸热效应是产生于在第一轮与第二轮加热期间样品在测试池外吸收的水分。第二轮曲线显示的一个大重结晶峰只在结晶不完全时(骤冷)才发生。

Interpretation The measurements show that the absorbed moisture evaporates relatively slowly from the approx. 1 mm thick sample. The TGA weight loss step and the endothermic DSC peak are correspondingly broad. The increasing mass loss above 300℃ is due to the onset of degradation of the polyamide.

The small endothermic event visible on the second DSC curve trace is due to water that has been absorbed by the sample outside the measuring cell between the first and second heating runs. The second curve shows a large recrystallization peak that only occurs when the crystallization is incomplete (quench-cooling).

结论 TGA 和 DSC 都可用来测定水分含量。两种技术的检测极限都为 0.1%。

再结晶峰表示结晶不完全。

Conclusions Both TGA and DSC can be used to determine the moisture content. The detection limit with both techniques is about 0.1%.

The recrystallization peak indicates that crystallization is incomplete.

不同加工批次的聚酰胺66/聚酰胺6
PA66/PA6 batches of different processability

样品	聚酰胺66 Ultramid C3ZM6 两批次（球状颗粒） 好的批次，12.0 mg；含单体的批次，5.9 mg		
Samples	Polyamide 66，2 batches of Ultramid C3ZM6 (pellets) Good batch，12.0 mg，batch with monomer，5.9 mg		
条件 **Conditions**	测试仪器：DSC	**Measuring cell**：DSC	
	坩埚： 40μl 标准铝坩埚，钻孔盖	**Pan**： Aluminum standard 40μl, pierced lid	
	样品制备： 从球状颗粒中心部分切出圆片	**Sample preparation**： Disk cut from center of pellet	
	DSC 测试： 以 10 K/min 从 30℃加热到 300℃	**DSC measurement**： Heating from 30℃ to 300℃ at 10 K/min	
	气氛：氮气，50 cm³/min	**Atmosphere**：Nitrogen，50 cm³/min	

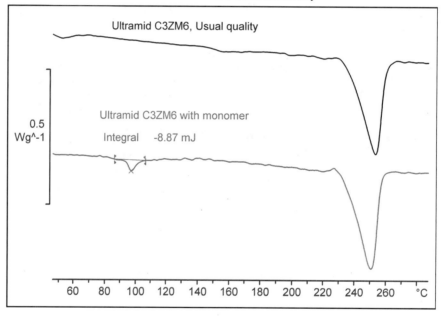

计算 小熔融峰积分给出的值为 8.9 mJ。峰的大小直接与单体含量成比例。设定 ε-己内酰胺的熔融热是 150 J/g，则单体含量的计算值为 1%：[8.9 mJ/(150 J/g · 5.9 mg)] · 100%。

Evaluation Integration of the small melting peak gave a value of 8.9 mJ. The size of the peak is directly proportional to the monomer content. If the heat of fusion of ε-caprolactam is assumed to be 150 J/g, a value of 1% is calculated for the monomer content：[8.9 mJ/(150 J/g · 5.9 mg)] · 100%.

解释 约100℃时的小吸热峰是由于单体(ε-己内酰胺)的熔融。含单体的批次会造成生产问题(在吹塑过程中会阻塞气孔)。好批次在100℃附近不会出现峰。

结论 单体含量大于0.2%时能在DSC熔融峰上检测出。TGA甚至会更灵敏,也适用于非结晶单体。通常,单体在加热时会蒸发。

Interpretation The small endothermic peak at about 100℃ is due to the melting of monomers (ε-caprolactam). The batch containing monomers caused manufacturing problems (clogging of the gas orifices in the blowing process). The good batch does not exhibit a peak around 100℃.

Conclusion Monomer contents greater than about 0.2% can be detected in the DSC melting peak. TGA would be even more sensitive and is also applicable to non-crystalline monomers. In general, monomers usually evaporate on heating.

错误认定的聚酰胺 6 和聚酰胺 66　PA6 and PA66, Mistaken Identity

样品	两批黑色手柄,据称是由 PA66 制成		
Samples	Two deliveries of black handles, said to be made of PA66		
条件	测试仪器:DSC	**Measuring cell**: DSC	
Conditions	坩埚:	**Pan**:	
	40 μl 标准铝坩埚,钻孔盖	Aluminum standard 40 μl, pierced lid	
	样品制备:	**Sample preparation**:	
	从手柄中心部分切出圆片,	Disk cut from the center of the handles.	
	货物 A:6.2 mg,货物 B:8.3 mg	Shipment A: 6.2 mg, shipment B: 8.3 mg	
	DSC 测试:	**DSC measurement**:	
	以 10 K/min 从 50℃加热到 300℃	Heating from 50℃ to 300℃ at 10 K/min	
	气氛:氮气,50 cm³/min	Atmosphere: Nitrogen, 50 cm³/min	

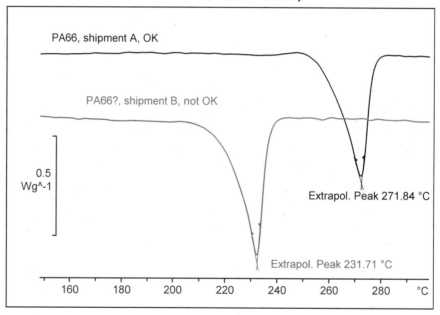

计算　在本特例中,检查峰温就足够了:
货物 A:271.8℃,因此是 PA66;货物 B:231.7℃,因此是 PA6。

Evaluation　In this particular case it was sufficient to check the peak temperatures:
Shipment A: 271.8℃, therefore PA66; shipment B: 231.7℃, therefore PA6.

解释　DSC 熔融峰立即显示只有货物 A 是由 PA66 制成的。高峰温表示它可能是 Bergamid A70B30。货物 B 明显是由 PA6 制成的,因此被拒收。

Interpretation　The DSC melting peaks immediately showed that only shipment A was made of PA66. The high peak temperature indicates that it could be Bergamid A70B30. Shipment B is clearly made of PA6 and was therefore rejected.

结论 不同类型的聚酰胺可以通过测试它们的 DSC 峰温来区分。用 DSC 检测原材料可以避免以后来自成品最终用户的申诉。使用较高加热速率如 20 K/min 可以将 DSC 测试所需的时间减至 10 分钟内。

Conclusions The different types of polyamide can be distinguished by measuring their DSC peak temperatures. Checking the raw materials by DSC avoids complaints later on from the end user of the finished product. The time needed for a DSC run could be reduced to less than 10 minutes by using a higher heating rate, e.g. 20 K/min.

聚酰胺 6/聚酰胺 66 共混物　PA6/PA66 Blend

样品 **Sample**	据称是由 PA6/PA66 共混物制成的安全帽 Protective caps said to be made of a PA6/PA66 blend	
条件 **Conditions**	测试仪器：DSC	Measuring cell：DSC
	坩埚： 40 μl 标准铝坩埚，钻孔盖	Pan： Aluminum standard 40 μl, pierced lid
	样品制备： 从帽上切下的小圆片(15.9 mg)	Sample preparation： Disk cut from cap (15.9 mg)
	DSC 测试： 以 10 K/min 从 30℃加热到 300℃	DSC measurement： Heating from 30℃ to 300℃ at 10 K/min
	气氛： 氮气，50 cm³/min	Atmosphere： Nitrogen，50 cm³/min

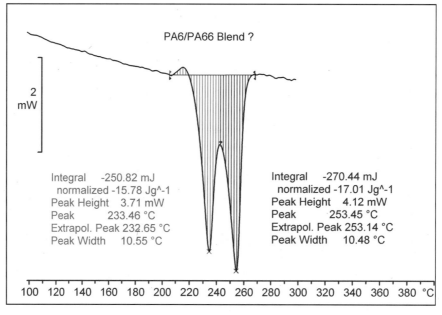

计算 Evaluation		第一个峰 First peak	第二个峰 Second peak
	峰温，℃ Peak temperature in ℃	233.5	253.4
	熔融热，J/g Heat of fusion in J/g	15.8	17.0

如果已知（或可以估算）相应的熔融热，则可以从峰面积估算出含量。如果假设两种材料的熔融热是 45 J/g（聚酰胺的典型值），则可计算出两种聚合物的含量为 35％和 38％。通过校准已知组成的共混物还能得到更准确的结果。

The contents can be estimated from the peak areas if the respective heats of fusion are known (or can be estimated). If one assumes that the heat of fusion for both materials is 45 J/g (typical for polyamides), contents of 35％ and 38％ are calculated for the two polymers. More accurate results could be obtained through calibration with blends of known composition.

解释 DSC 曲线显示两个表明存在 PA6 和 PA66 的分离的熔融峰。为了获得原始熔融热,基线开始于小放热再结晶峰前(如所应用的)。

结论 安全帽实际上是由 PA6/PA66 共混物制成的。峰面积能对这两种聚合物进行半定量分析。

Interpretation The DSC curve shows two separate melting peaks that indicate the presence of PA6 and PA66. The baseline is drawn from before the small exothermic recrystallization peak in order to obtain the original heat of fusion (as supplied).

Conclusions The protective caps are in fact made of a PA6/PA66 blend. The peak areas allow a semiquantitative analysis of the two polymers to be made.

聚酰胺 6 共混物　Polyamide 6 Blends

样品　Triax RDS 13415 球状颗粒（ABS,聚丙烯腈-丁二烯-苯乙烯与聚酰胺 6 的共混物）
Flexovac-Multi（PE 和 PA6 的层压薄膜）

Samples　Triax RDS 13415 pellets (a blend of ABS, polyacrylonitrile-butadiene-styrene, with polyamide 6)
Flexovac-Multi (laminated film of PE and PA6)

条件

测试仪器：DSC

坩埚：
40 μl 标准铝坩埚，钻孔盖

样品制备：
从球状颗粒切出和从薄膜冲出的圆片

样品质量：
ABS/PA6 共混物 9.4 mg，Flexovac 9.0 mg

DSC 测试：
以 20 K/min 从 30℃加热到 280℃（Trias），以 10K/min 从 50℃加热到 300℃（Flexovac）

气氛：氮气，50 cm³/min

Conditions

Measuring cell：DSC

Pan：
Aluminum standard 40 μl, pierced lid

Sample preparation：
Disk cut from pellet and punched out of film

Sample mass：
ABS/PA6 blend 9.4 mg, Flexovac 9.0 mg

DSC measurement：
Heating from 30℃ to 280℃ at 20 K/min (Triax), and from 50℃ to 300℃ at 10 K/min (Flexovac)

Atmosphere：Nitrogen, 50 cm³/min

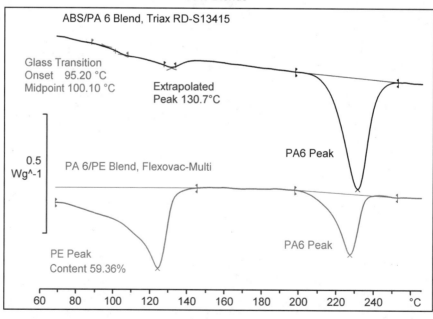

计算　PA6 的熔融热约为 45 J/g（PA6 的典型值）。两个样品得到的值都为 20 J/g。因此它们的 PA6 含量约为 44%，假设它们的结晶度相同。如果事前已经输入了熔融热

Evaluation　The heat of fusion of PA6 is approx. 45 J/g (typical for PA6). Both samples give values of 20 J/g. Their PA6 content is therefore approx. 44% assuming that their crystallinity is the same. The calculation of the content is performed automatically if the reference value for the heat of

参比值,含量的计算则会自动进行。用相似的方法,Flexovac 的 PE 含量确定为 59%。

解释　如同预期的,在两条曲线的约 230 ℃处都可以看到 PA6 熔融峰。Triax RDS 的 ABS 成分使得 100℃处出现聚苯乙烯的玻璃化转变,130℃处出现聚丙烯腈的玻璃化转变。Flexovac-Multi 的 PE 成分显示在 124℃的熔融峰(相当于 100J/g 的熔融热)。

结论　DSC 是用于鉴定这样的黑色材料的极好方法。用红外光谱来鉴定是困难的,因为黑色材料完全吸收射入的红外光。

fusion is entered beforehand. In a similar way, the PE content of Flexovac is determined to be 59%.

Interpretation　As expected, the PA6 melting peak can be seen on both curves at about 230℃. The ABS component of Triax RDS gives rise to glass transitions of polystyrene at 100℃ and of polyacrylonitrile at 130℃. The PE component of Flexovac-Multi exhibits a melting peak temperature of 124℃ (corresponding to a heat of fusion of about 100 J/g).

Conclusion　DSC is an excellent method to use for the identification of such black materials. Identification by infrared spectroscopy is difficult because black materials completely absorb the incident IR light.

用 IsoStep DSC 测定聚酰胺 6 的玻璃化转变和水分含量
Glass Transition and Moisture Content of PA6 by IsoStep DSC

样品　聚酰胺薄膜，PA6。实验目的是研究水分对玻璃化转变的影响。为此，一部分薄膜被真空干燥，并存放在硅胶上（样品名为"干的"）；另一部分浸入水中（样品名为"饱和的"）。为了得到中等水分含量，将饱和样品保持在相对湿度约为 60% 的空气中 4 小时（样品名为"湿的"）。

Sample　Polyamide film, PA6. The aim of the experiment was to investigate the influence of moisture on the glass transition. To do this, one part of the film was vacuum-dried and stored over silica gel (sample named "dry"); the other part was immersed in water (sample named "saturated"). To obtain a sample of intermediate moisture content, the saturated sample was kept in air at approx. 60% relative humidity for 4 hours (sample named "moist").

条件　测试仪器：DSC
Conditions

坩埚：
带密封盖和钻孔（1mm 孔）盖的 40μl 铝坩埚（用于常规 DSC）；带盖 20μl 铝坩埚用于 IsoStep 测试

样品制备：
样品被切成一些小片。样品质量约为 4 mg。在制备样品前，饱和样品的多余水分被完全除去。

DSC 测试：
常规程序：以 10 K/min 的升温速率从 −60℃ 加热到 130℃。
IsoStep 程序：
从 −60℃ 阶梯式加热到 110℃，等温周期为 60 s，加热段为 2K/min，温度增量为 2℃。在相同的条件下进行空白与蓝宝石参比样品测试以进行校准。

IsoStep 是一种特殊的 DSC 测试技术。它能将比热变化从反应或蒸发这样的重叠动态效应（也称为不可逆效应）中分离开。这是通过一种含有大量都以等温段开始和结束的动态段组成的温度程序来实现的。专用的 IsoStep 数据处理得到消除了重叠蒸发过程的比热（c_p）曲线。

Measuring cell: DSC

Pan:
Aluminum 40μl with sealed and pierced (1 mm hole) lids (for conventional DSC); aluminum 20μl with lid for IsoStep measurements.

Sample preparation:
Samples were cut into a few small pieces. Sample mass was about 4 mg. The excess water of the saturated sample was completely removed before preparing the specimen.

DSC measurement:
Conventional program: Heating from −60℃ to 130℃ at a heating rate of 10K/min.
IsoStep program:
Stepwise heating from −60℃ to 110℃ with isothermal periods of 60 s and heating segments of 2 K/min, with temperature increments of 2℃. The blank and sapphire reference samples were measured under the same conditions for calib-ration purposes.

IsoStep is a special DSC measurement technique. It allows a change in heat capacity to be separated from an overlapping kinetic effect such as a reaction or evaporation (also called a non-reversing effect). This is done by means of a temperature program consisting of a large number of dynamic segments that each begin and end with an isothermal segment. The dedicated IsoStep evaluation yields the specific heat

不可逆曲线是蒸发热流。

气氛：氮气，50 ml/min

capacity (C_p) curve free from the overlapping evaporation process. The non-reversing curve is the heat flow of the evaporation.

Atmosphere：Nitrogen，50 ml/min

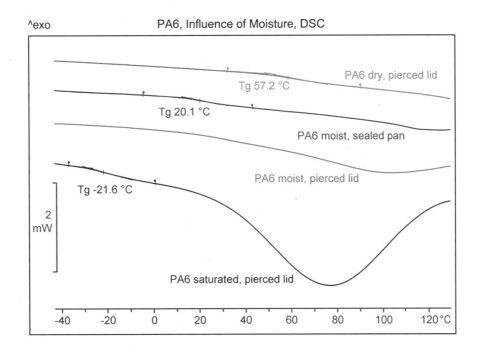

解释 对不同水分含量(干的、湿的和饱和的)的 PA6 常规测试的 DSC 曲线显示一个或两个热效应，即玻璃化转变台阶和蒸发吸热峰。随着水分含量的增加，PA6 薄膜的玻璃化转变温度降低。干薄膜显示玻璃化温度(T_g)为 57℃，饱和薄膜为 −22℃。因此水起着增塑剂的作用，取决于它的浓度，薄膜在室温下是硬的或软的。

如果坩埚是完全密封的，水分会保留在样品中；高达 100℃ 也没有测出蒸发。在更高温度时，内压变大，可能会冲开坩埚。要承受住增加的压力，应使用压力坩埚进行测试。然而，这样会造成较差的灵敏度。

如果用钻孔盖来封住坩埚，水分在加热时蒸发，峰大小与样品中存在的水分量相对应，例如含水 6.9% 的饱和薄膜。蒸发使得测定玻璃化转变变得困难。如果蒸发发生在相

Interpretation The conventionally measured DSC curves of PA6 films with different moisture contents (dry, moist and saturated) display one or two thermal effects, namely the step at the glass transition and the endothermic evaporation peak. With increasing moisture content, the glass transition temperature of the PA6 film decreases. The dry film exhibits a glass transition temperature (T_g) at 57℃, and the saturated film at −22℃. The water therefore acts as a plasticizer. Depending on its concentration, the film can be either stiff or soft at room temperature.

The moisture is held in the sample if the pan is hermetically sealed; no evaporation is detected up to 100℃. At higher temperatures, the internal pressure becomes large and might rupture the pan. To withstand the increased pressure, measurements would have to be performed with pressure pans. These, however, result in poorer sensitivity.

If the pan is sealed with a pierced lid, moisture evaporates during heating and the peak size corresponds to the amount of moisture present in the sample, for example 6.9% for the saturated film. This evaporation makes it difficult to determine the glass transition. If evaporation takes place at the same temperature,

同的温度,就会因为效应重叠而无法进行数据计算。但是在样品制备期间,PA6 也可能在空气中变干,玻璃化转变发生移动,如湿的 PA6 所示的两条曲线。

为了更容易地识别玻璃化转变,使用 IsoStep 方法来将玻璃化转变效应与水分蒸发效应分开。这能准确地确定 T_g;水分含量可以从蒸发峰或失重测定。

the evaluation becomes impossible because the effects overlap. But also during sample preparation, the PA6 may undergo drying in air, and the glass transition is shifted as the two curves for the moist PA6 show.

In order to identify the glass transition more easily, the IsoStep method was used to separate effect of the glass transition from that of the evaporation of moisture. This allows T_g to be accurately assigned; the amount of moisture can be measured from the evaporation peak or by the weight loss.

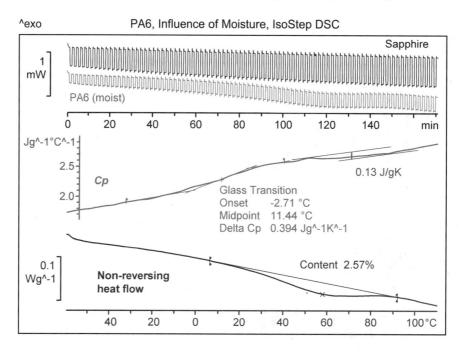

图的上半部分显示典型的以时间为函数的 IsoStep 测试曲线。这些曲线是空白校正过的蓝宝石曲线(空白)与样品曲线(红色)。这两条曲线的 IsoStep 数据处理计算了 C_p 曲线(绿色)与不可逆热流曲线(蓝色)。

比热曲线显示玻璃化转变为 11℃。水分蒸发产生一个不可逆曲线上的宽吸热峰。这伴随着因为失水造成的约 0.13 J/gK 的热容下降。对不可逆曲线上的放热峰积分,计算得到样品中水分含量约为 2.6%(基于水的蒸发热为 2400 J/g)。该结果与样品失重得到的吻合。该值也与因水分损失造成的热容降低(约 0.13 J/gK)相符。

The upper part of the diagram displays the typical IsoStep measurement curves as a function of time. These curves are the blank-corrected sapphire curve (black) and the sample curve (red). The IsoStep evaluation of these two curves calculates the C_p curve (green) and the non-reversing heat flow curve (blue).

The curve of the specific heat capacity shows a glass transition at 11℃. The evaporation of moisture gives rise to a broad endothermic peak on the non-reversing curve. This is associated with a decrease in heat capacity of about 0.13 J/gK due to the loss of water. Through integration of the exothermic peak in the non-reversing curve, the water content in the sample was calculated to be about 2.6% (based on a heat of evaporation of water of 2400 J/g). The result agrees well with that obtained from the weight loss of the sample. The value also matches the decrease in heat capacity (about 0.13 J/gK) due to the water lost.

结论 水在聚酰胺样品中起着增塑剂的作用,使玻璃化转变温度从干燥条件下的 57℃ 移至水饱和 PA6 样品的 -22℃。这些含过量水分样品的 T_g 用常规 DSC 实验可容易地测定。

如果水分含量足够大致使密封坩埚中的压力显著上升,困难就会出现。坩埚会变形或被冲破。如果使用钻孔盖坩埚,蒸发会与玻璃化转变重叠。在这种情况下,IsoStep 方法能分开测定热容变化和蒸发热流。这样,即使因为热效应重叠使蒸发掩盖了玻璃化转变,也能确定湿样品的 T_g。

Conclusions Moisture acts as a plasticizer in the polyamide samples and shifts the glass transition temperature from 57℃ under dry conditions to 22℃ for the water-saturated PA6 sample. The T_g of these samples with extreme moisture contents can be easily determined by conventional DSC experiments.

Difficulties arise if the moisture content is sufficiently large to cause a significant rise in pressure in the sealed pan. This can then deform or rupture it. If pans with pierced lids are used, the evaporation overlaps the glass transition. In such cases, the IsoStep method is able to determine the heat capacity changes and the heat flow of the evaporation separately. This allows the T_g of moist samples to be identified even if the evaporation masks the transition because the thermal effects overlap.

7.6 聚对苯二甲酸乙二醇酯的热行为
Thermal behavior of polyethylene terephthalate

聚对苯二甲酸乙二醇酯的热历史　PET, Thermal history

样品 Sample	取自软饮料瓶颈的聚对苯二甲酸乙二醇酯 Polyethylene terephthalate from the neck of a bottle for soft drinks	$[-O-(CH_2)_2-O-\overset{O}{\underset{\|\|}{C}}-\underset{}{\bigcirc}-\overset{O}{\underset{\|\|}{C}}-]_n$
条件 Conditions	测试仪器:DSC 坩埚: 40 μl 标准铝坩埚,钻孔盖 样品制备: 从瓶颈切下的扁平片,23.16 mg DSC测试: 以 10 K/min 从 30℃ 加热到 300℃。在 300℃ 时,自动进样器在几秒钟内取出熔融样品,将坩埚放在冷转盘上。这会产生平均冷却速率约为 50 K/s 即 3000 K/min 的可重复性骤冷。 当测试池达到 30℃ 时,放入坩埚,以 10 K/min 再次测试。骤冷后,为使焓松弛在 65℃ 退火 10 小时。然后以 10 K/min 从 30℃ 到 300℃ 再测试样品。最后以 10 K/min 冷却以能有时间结晶。最后,以 10 K/min 再次测试。 气氛:氮气,50 cm³/min	Measuring cell:DSC Pan: Aluminum standard 40 μl, pierced lid Sample preparation: Flat piece cut from neck of bottle, 23.16 mg DSC measurement: Heating from 30℃ to 300℃ at 10 K/min. At 300℃ the sample changer removes the molten sample within a few seconds and places the pan on the cold tray. This results in reproducible quench-cooling with a mean cooling rate of approx. 50 K/s or 3000 K/min. As soon as the cell reaches 30℃ the pan is inserted and measured again at 10 K/min. After quench-cooling, it is annealed at 65℃ for 10 h for enthalpy relaxation. The sample is then measured again from 30℃ to 300℃ at 10 K/min. The final cooling is at 10 K/min to allow time for crystallization. Finally, it is measured again at 10 K/min. Atmosphere:Nitrogen, 50 cm³/min
计算 Evaluation		

样品 Sample	骤冷 Quench-cooled	退火 Annealed	缓慢冷却 Cooled slowly
T_g,起始点,℃ T_g, onset in ℃	76.2	82.7	75.6
Δc_p,J/gK	0.37	0.32	0.20
结晶热,J/g Heat of crystallization in J/g	35.4	34.3	0.1
熔融热,J/g Heat of fusion in J/g	42.3	39.9	39.9
熔融峰温,℃ Melting peak temperature in ℃	248.6	248.6	248.6

使用直线基线在 120℃ 到 180℃ 间对结晶热进行积分,熔融热在 210℃ 到 270℃ 间。为了清晰起见,计算结果没有显示在图中。

The heats of crystallization have been integrated using a straight baseline between 120℃ and 180℃, and the heats of fusion between 210℃ and 270℃. For the sake of clarity, the evaluation results are not shown in the diagram.

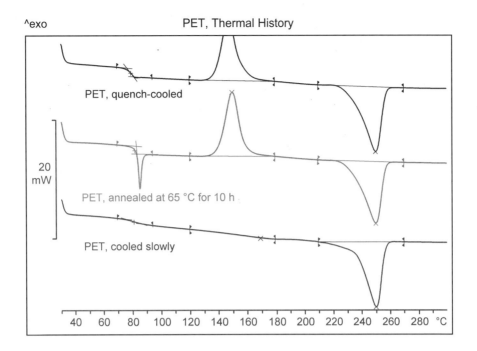

解释 骤冷样品（黑色曲线）通常在玻璃化转变区域显示曲线的台阶状变化。另一方面，65 ℃的退火产生松驰峰（红色曲线上）。缓慢冷却的样品的玻璃化转变（蓝色曲线）不明显，因为部分聚合物此时已经处于结晶态；产生的比热变化只有一半值。

部分结晶的样品在 T_g 以上经历冷结晶。在这种情况下，测定测试开始时的原始结晶度很困难。用 ($\Delta H_{fus} - \Delta H_{cryst}$) 计算整个焓变对测定原始结晶度是关键，因为熔融焓是温度依赖的，基线的构建可能会不正确。

结论 许多材料的性能受它们热历史的影响，例如，金属玻璃、硫和硒元素，和乙酰水杨酸或乙醇（低于室温测试）这样的简单有机化合物。玻璃化转变温度和熔融温度当然各不相同，因为当样品从熔融态骤冷时，冷却速率对防止结晶来说是绝对重要的。

第二轮加热应该用来测定消除了未知热历史或机械历史影响的无定形材料的性质。为此，第一轮测试的

Interpretation The quench-cooled sample (black curve) as usual exhibits the step-like change of the curve in the glass transition region. On the other hand, annealing at 65℃ gives rise to a relaxation peak (on the red curve). The glass transition of the sample that was slowly cooled (blue curve) is less pronounced because part of the polymer is now in a crystalline state; the resulting heat capacity change is about half the value.

Partially crystalline samples undergo cold crystallization above the T_g. In such cases, the determination of the original crystallinity at the beginning of the measurement is difficult. The calculation of the overall enthalpy change by ($\Delta H_{fus} - \Delta H_{cryst}$) for determination of the initial crystallinity can be critical because the enthalpy of fusion is temperature dependent and the construction of the baseline may be incorrect.

Conclusions The properties of many materials are influenced by their thermal history, e. g. metallic glasses, the elements sulfur and selenium and simple organic compounds such as acetylsalicylic acid or ethanol (subambient measurement). The glass transition temperatures and melting temperatures are, of course, individually different as are the cooling rates necessary to prevent crystallization when the samples are quench-cooled from the melt.

The second heating run should be use to determine the properties of amorphous materials free from the influence of unknown thermal or mechanical history. To do this, the sample from the

样品从熔融态被骤冷。另外，如果要研究感兴趣的最大结晶度，熔融态样品应该非常缓慢地冷却（最大速率 10 K/min）。

上述比较复杂的温度程序步骤可用自动进样器自动进行。

聚对苯二甲酸乙二醇酯的焓松驰　PET, Enthalpy Relaxation

样品 **Sample**	取自软饮料瓶颈的聚对苯二甲酸乙二醇酯 Polyethylene terephthalate from the neck of a soft drinks bottle	
条件 **Conditions**	测试仪器：DSC	**Measuring cell**：DSC
	坩埚：40μl 标准铝坩埚，钻孔盖	**Pan**：Aluminum standard 40μl, pierced lid
	样品制备：	**Sample preparation**：
	从瓶颈切下的扁平片，23.16 mg	Flat piece cut from the neck of the bottle, 23.16 mg
	DSC 测试：	**DSC measurement**
	以 10 K/min 从 30 加热到 300 ℃。在 300 ℃时，自动进样器在几秒钟内取出熔融样品，将坩埚放在冷转盘上。这会产生平均冷却速率约为 3000 K/min 的可重复性骤冷。	Heating from 30℃ to 300 ℃ at 10 K/min. At 300 ℃ the sample changer removes the molten sample within a few seconds and places the pan on the cold tray. This results in reproducible quench-cooling with a mean cooling rate of approx. 3000 K/min.
	骤冷后，样品在 65℃下退火 0 至 24 小时使焓松驰，每次再以 10 K/min 从 30℃加热到 300℃进行测试。整个温度程序可用自动进样器在周末时全自动完成。	After quench-cooling, the sample was annealed at 65 ℃ for periods of 0 to 24 h for enthalpy relaxation and each time measured again from 30℃ to 300℃ at 10 K/min. The entire temperature program was executed fully automatically using the sample changer over a weekend.
	气氛：氮气，50 cm³/min	**Atmosphere**： Nitrogen, 50 cm³/min

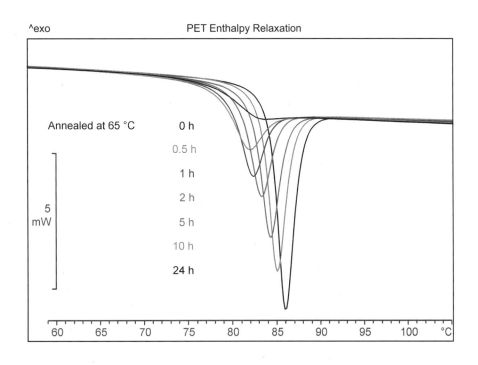

计算
Evaluation

65 ℃下的退火时间(小时): Annealing at 65℃ in h:	0	0.5	1	2	5	10	24
峰面积,J/g Peak area in J/g	—	2.0	2.6	3.2	4.0	4.6	5.3
T_g,起始点,℃ T_g, onset, ℃	77.8	78.1	79.2	80.3	81.7	82.7	83.7
T_g,中点,℃ T_g, midpoint, ℃	80.0	78.5	78.8	79.7	80.8	81.6	82.7
T_g,拐点,℃ T_g, inflect. Point in ℃	80.7	80.3	80.9	81.9	83.0	83.9	84.7

注意:如果发生相当大的焓松驰,则由中点确定的玻璃化转变温度会比由起始点确定的低(见下图)。

Note: The glass transition temperature determined by the midpoint, can be lower than that by the onset if a rather large enthalpy relaxation occurs (see next diagram).

解释 第一张图(PET Enthalpy Relaxation)显示峰的大小随着退火时间的增加而增加。吸热峰正完全对应于在玻璃化温度以下退火时由于松驰导致的焓的减少。

在第二张图(PET, Evaluating Glass Transition)中,对两条曲线作为示例进行计算,计算区间为73℃和92℃。

Interpretation The first diagram (PET Enthalpy Relaxation) shows that the peak size increases with increasing annealing time. This endothermic peak corresponds exactly to the enthalpy decrease due to relaxation during annealing below the glass transition.

In the second diagram (PET, Evaluating Glass Transition), two curves are evaluated as examples with evaluation limits of 73℃ and 92℃.

结论 焓峰影响了对玻璃化转变的计算,它通常是热历史的结果。为了消除热历史,对研究中的样品加热至略高于玻璃化转变的温度,骤

Conclusions The enthalpy peak influences the evaluation of the glass transition and is often a result of the thermal history. To eliminate thermal history, the sample under investigation is heated to a temperature slightly above the glass transition,

冷,然后第二次测试。结晶较容易的聚合物(例如 PET)被加热至熔点以上,然后骤冷以遏制微晶的形成(只有能运动的无定形部分才有比热变化)。

不过,松驰焓能提供关于样品热和机械历史的重要信息(例如,贮存温度、贮存时间、冷却速率)。

quench-cooled, and then measured a second time. Polymers that crystallize readily (such as PET) are heated to above the melting point and quench-cooled to suppress the formation of crystallites (only the mobile amorphous fraction undergoes a change in heat capacity).

The relaxation enthalpy can, however, provide important information regarding the thermal and mechanical history of a sample (e. g. storage temperature, storage time, cooling rate).

用动态负载 TMA 测定聚对苯二甲酸乙二醇酯的冷结晶
PET, Cold Crystallization by Dynamic Load TMA

样品 聚对苯二甲酸乙二醇酯圆片(梅特勒-托利多教学样品包)
Sample Disk of polyethylene terephthalate (METTLER TOLEDO tutorial sample kit)

条件 测试仪器:TMA,3mm 圆点探头
Conditions **Measuring cell**:TMA with 3mm ball-point probe

样品制备:
样品放在直径 6mm、厚 0.5mm 的石英圆片上

Sample preparation:
The sample is placed on a fused silica disk of 6-mm diameter and 0.5mm thickness

TMA 测试:
以 10 K/min 从 25℃加热到 150℃
负载:每 6 秒从 0.01 N 变到 0.19 N
气氛:静态空气

TMA measurement:
Heating from 25℃ to 150 ℃ at 10 K/min
Load:Changing every 6 s from 0.01 to 0.19 N
Atmosphere　Static air

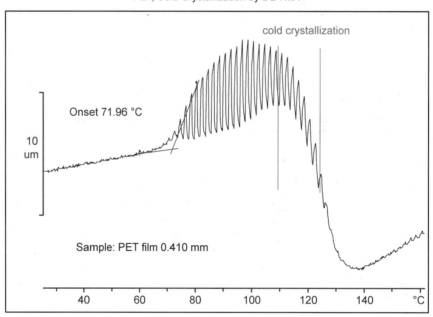

计算 振幅增加的起始点 71.96℃等于玻璃化转变温度。

Evaluation The onset of the amplitude increase of 69.2 ℃ corresponds to the glass transition temperature.

解释　低的 0.10N 力平均值在 70℃左右的玻璃化转变处没有产生可见的凹陷。相反,由于先前取向的膜的应力释放,样品轻微变厚。在玻璃化转变以上,样品对周期性力的变化发生粘弹性响应。在 95℃到达最大振幅。继续加热时,样品由于冷结晶发生约 4.3%的收缩。振幅减小。在 140℃,结晶过程实际上结束了,样品又变硬。与化学反应不同,结晶可以通过融化微晶反向进行。

结论　动态负载 TMA 可以用来研究 PET 相当特殊的热机械性能。DSC 当然也能测试玻璃化转变和冷结晶(见应用实例"聚对苯二甲酸乙二醇酯的热历史"),但是 DSC 曲线不能提供任何力学性能的信息。

Interpretation　The low mean force of 0.10 N causes no visible indentation at the glass transition around 70 ℃. On the contrary, the sample becomes slightly thicker due to stress relief of the previously oriented film. Above the glass transition, the sample responds viscoelastically to the periodical force change. The maximum amplitude is reached at 95 ℃. On further heating, the sample undergoes shrinkage of approx. 4.3% due to crystallization. The amplitude decreases. At 140 ℃, the crystallization process is practically finished and the sample is again hard. In contrast to chemical reactions, the crystallization can be reversed by melting the crystallites.

Conclusions　Dynamic load TMA can be used to investigate the rather special thermomechanical properties of PET. DSC can of course also measure the glass transition and cold crystallization (see Application Example "PET, Thermal History"), but the DSC curves do not provide any information on mechanical behavior.

聚对苯二甲酸乙二醇酯的动态热机械分析 Dynamic Mechanical Analysis of PET

样品　三种不同的PET薄膜：样品A缓慢结晶；样品B为所谓的低粘度；样品C为标准粘度。为表征物理转变和比较熔融和结晶时的行为而进行测试。

Sample　Three different PET films; sample A crystallizes slowly; sample B has so-called low viscosity and sample C has standard viscosity.
The measurements were performed to characterize the physical transitions and to compare behavior during melting and crystallization.

条件　仪器：
Conditions　DMA861^e，剪切夹具

样品制备
两片厚0.94mm、直径5mm的圆片装在剪切样品夹具中。

DMA测试：
1) 剪切测试在1 Hz下进行，以2K/min的速率从−140℃加热到280℃。最大力振幅为25N；最大位移振幅为5μm；偏移控制为零。

2) 对比测试在10 Hz下进行，以2K/min的速率从25℃加热到280℃，接着2K/min冷却到100℃；频率为10Hz；温度范围为25℃到280℃；最大力振幅为3N；最大位移振幅为30μm；偏移控制为零。

气氛：静态空气

Measuring cell：
DMA/SDTA 861^e with the shear sample holder

Sample preparation：
Two disks, 0.94mm thick of 5mm diameter were mounted in the shear sample holder.

DMA measurement：
1) The shear measurement was performed at 1 Hz from −140℃ to 280 ℃ at a heating rate of 2 K/min. Maximum force amplitude 25 N; maximum displacement amplitude 5 μm; offset control zero.

2) The comparison measurements were performed at 10 Hz from 25℃ to 280 ℃ at a heating rate of 2 K/min followed by cooling to 100 ℃ at 2 K/min. Maximum force amplitude 3 N; maximum displacement amplitude 30 μm; offset control zero.

Atmosphere：Static air

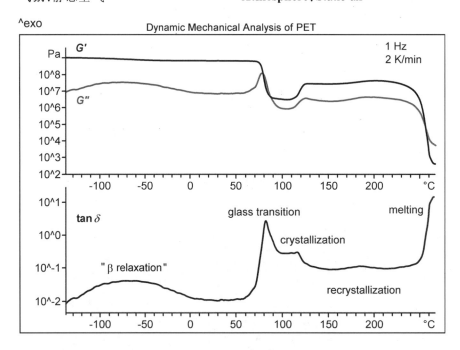

解释 PET 样品 A 用 DMA 程序 1 测试。第一张图中的曲线显示整个温度范围内 PET 的行为。在室温以下，可观察到次级松弛。此 β 松弛与分子链段的小尺寸运动相对应。该效应与在 80℃ 观察到的主松弛（玻璃化转变）相比是小的。由于在约 120℃ 冷结晶，导致储能模量提高。继续加热时，发生再结晶和微晶熔化。熔融过程中，储能模量急剧下降，G' 显示从 -140℃ 的 10^9 Pa 下降到 270℃ 的 $5 \cdot 10^2$ Pa，总的变化为 6.5 个数量级。

Interpretation The PET sample A was measured using DMA program 1. The curves in the first diagram show the behavior of PET over the whole temperature range. At subambient temperatures, the secondary relaxation can be observed. This β relaxation corresponds to small-scale movements of molecular segments. This effect is small in compa- rison to the main relaxation (glass transition) observed at 80℃. The storage modulus increases due to cold crystallization at about 120℃. On further heating, recrystallization and melting of the crystallites takes place. During melting, the storage modulus decreases dramatically so that G' shows an overall change of 6.5 decades from 109 Pa at -140℃ to 5 102 Pa at 270℃.

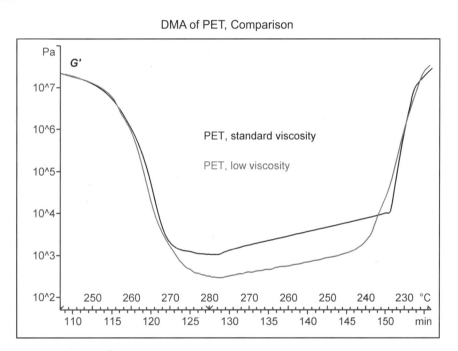

加热和冷却实验显示两个 PET 材料 B 和 C 的熔融和结晶行为（用程序 2 进行 DMA 测试）。熔融期间稳定的模量信号用 30μm 的位移振幅得到，该位移振幅对于剪切测试是相当高的。两个熔融样品的模量差约一个数量级，这对于材料的加工是重要的，因此这两种材料被标记为"标准"或"低"粘度。

Heating and cooling experiments show the melting and crystallization behavior of the two PET materials B and C (DMA measurement using program 2). The stable modulus signal during the melt is obtained using a displacement amplitude of 30μm, which is rather high for shear measurements. The moduli of the molten samples differ by about one decade. This is of importance for the processing of the materials; the materials are labeled accordingly as being of "standard" or "low" viscosity.

结论 测试显示 PET 这样的聚合物可以容易地用 DMA/SDTA861^e 测试，研究从半结晶或坚硬无定形状态

Conclusions The measurements show that a polymer such as PET can be easily measured with the DMA/SDTA861^e to study dynamic mechanical and rheological behavior from the

到低粘度熔体状态的动态力学和流变行为。这些信息可以用来优化作特殊用途的聚合物材料的质量。

semicrystalline or rigid amorphous state to the low viscous melt. The information can be used to optimize the quality of polymeric materials for their specific applications.

7.7 其他聚合物的测试 Measurements on other polymers

聚甲基丙烯酸甲酯的玻璃化转变 PMMA, Glass Transition

样品	有机玻璃板,4mm 厚（聚甲基丙烯酸甲酯 PMMA）	
Sample	Plexiglas pane, 4 mm thick (polymethylmethacrylate, PMMA)	
条件	测试仪器：DSC	**Measuring cell**：DSC
Conditions	坩埚：40 μl 标准铝坩锅,打孔盖	**Pan**：Aluminum standard 40 μl, pierced lid
	样品制备：	**Sample preparation**：
	从板上锯出圆片,用小刀削成合适的片,15.4 mg	Disks sawn from the pane and suitable pieces cut off with a knife, 15.4 mg
	DSC 测试：	**DSC measurement**：
	第一轮以 10K/min 从 40℃ 加热到 200℃,以 5K/min 冷却	First run from 40℃ to 200℃ at 10 K/min, cooling at 5 K/min, second run from 40℃ to 200℃ at 10 K/min
	第二轮以 10K/min 从 40℃ 加热到 200℃	
	气氛：氮气,50 cm³/min	**Atmosphere**：Nitrogen, 50 cm³/min

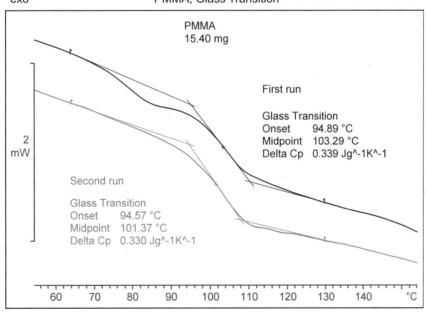

计算 玻璃化转变温度的中点为 103.3℃（第一轮）和 101.4℃（第二轮）。

Evaluation The midpoints of the glass transition temperatures are 103.3 ℃ (first run) and 101.4 ℃ (second run).

解释 PMMA 是完全无定形的,玻璃化转变温度约为 105℃。第一轮测试曲线几乎总是受松弛现象的影响。如果热历史被消除,DSC 曲线的形状会基本上变得理想。在上面所示的曲线中,松弛峰出现在实际玻璃化转变之前。不过,它也可能直接与玻璃化转变区域重叠或者在之后出现(见 PET 例)。

结论 对于 PMMA,为了获得可靠值通常最好是进行第二轮测试。

Interpretation PMMA is completely amorphous and has a glass transition tempe- rature of approx. 105℃. The first measurement curve is nearly always influenced by relaxation phenomena. If the thermal history is eliminated, the shape of the DSC curve becomes more or less ideal. In the curve shown above, the relaxation peak appears before the actual glass transition. It can however also overlap the glass transition region directly or appear afterward (see for example PET)

Conclusion With PMMA it is often advisable to perform a second measurement run in order to obtain a reliable values.

聚甲醛的 DSC 测试 DSC measurement of POM

样品 Hostaform C52021 球状颗粒 $(—CH_2—O—)_n$
Sample Hostaform C52021 as pellets

条件 测试仪器: DSC **Measuring cell**: DSC
Conditions 坩埚: 标准 40 μl 铝坩锅，打孔盖 **Pan**: Aluminum standard 40 μl, pierced lid
样品制备: 从球状颗粒切出圆片 **Sample preparation**: Disk cut from pellet
DSC 测试: **DSC measurement**:
以 10 K/min 从 30℃ 加热到 300℃， Heating from 30℃ to 300℃ at 10 K/min
气氛: 氮气, 50 cm^3/min **Atmosphere**: Nitrogen, 50 cm^3/min

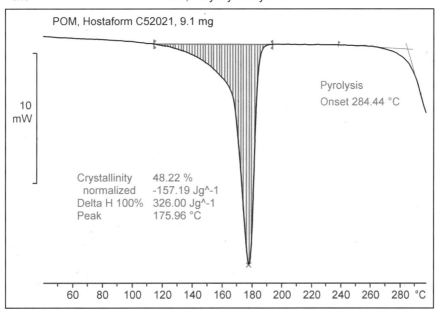

结晶度, % Degree of crystallinity in %	48.2
峰温, ℃ Peak temperature in ℃	176.0
热解起始点, ℃ Onset of pyrolysis in ℃	284.4

计算
Evaluation

解释 该熔融曲线实际上是 POM 均聚物所特有的(共聚物约低 10 K 熔融)。然而，供应商说所测试的样品是共聚物。260℃ 以上的吸热效应产生于热解开始。

Interpretation The melting curve is actually characteristic of a POM homopolymer (copolymers melt about 10 K lower). The supplier, however, said that the sample measured was a copolymer. The endothermic effect above 260 ℃ is caused by the onset of pyrolysis.

结论 聚甲醛是一种高结晶度的材料。热解开始于微晶熔程约 50 K 以上，伴随着甲醛的形成。这意味着如果样品在耐压坩锅中加热，将会有爆炸的危险。

Conclusions Polyoxymethylene is a material with a high degree of crystallinity. Pyrolysis begins about 50 K above the crystalline melting range with the formation of formaldehyde. This means that if the sample is heated in a pressure resistant crucible, there is the risk of an explosion.

聚乙二酸丙二醇酯的 DSC 测试　　DSC Measurements of PPA

样品　　PPA Amodel 注射成型部件，标准品和含 75％无机填料
Sample　PPA，Amodel，injection-molded parts，normal and with 75％ inorganic filler

条件　　仪器：DSC　　　　　　　　　　　　　　**Measuring cell**：DSC
Conditions　坩埚：标准 40μl 铝坩锅，盖钻孔　　　**Pan**：Aluminum standard 40μl, pierced lid
　　　　　　样品制备：从部件上切出圆片　　　　**Sample preparation**：Disk cut from part
　　　　　　DSC 测试：　　　　　　　　　　　**DSC measurement**：
　　　　　　第一轮加热以 20 K/min 从 35℃到　　First heating run from 35℃ to 350 ℃ at 20 K/
　　　　　　350℃。然后样品在自动进样器上　　min. The sample was then quench-cooled to room
　　　　　　骤冷至室温，在相同条件下进行第　temperature in the sample changer and a second
　　　　　　二轮测试。　　　　　　　　　　　　heating run performed under the same conditions
　　　　　　气氛：氮气，50 cm³/min　　　　　　**Atmosphere**：Nitrogen，50 cm³/min

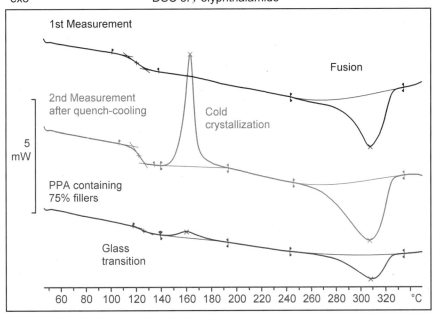

计算
Evaluation

	第一轮 First run	第二轮 Seconod run	含填料的 Filled
T_g，中点，℃ T_g, midpoint in ℃	120.8	122.3	126.0
结晶热，J/g Heat of crystallization in J/g	—	22.1	1.5
结晶峰温，℃ Crystallization peak temperature in ℃	—	163.0	160.5
熔融热，J/g Heat of fusion in J/g	30.7	37.6	8.2
熔融峰温，℃ Melting peak temperature in ℃	307.0	308.0	308.0
样品质量，mg Sample mass in mg	6.22	7.27	11.1

熔融峰用样条曲线为基线积分。　　　　　　　　The melting peaks were integrated using the baseline type spline.

解释 在所供给的注射成型部件中（第一轮测试），PPA 是半结晶的，玻璃化转变处的 DSC 曲线台阶高度小。事实上没有冷结晶。骤冷的样品是无定形的，显示清晰的玻璃化转变和冷结晶。

在填充材料中，所有测得的热效应更小，因为它们只与聚合物的含量有关。聚合物含量可以通过填充材料的熔融热和未填充材料的熔融热根据下式计算：

此例，填充材料的聚合物含量估算为 22%。

结论 DSC 曲线的分析提供有关热历史、结晶度和聚合物含量的信息。

Interpretation In the injection-molded part as supplied (first measurement), the PPA is semicrystalline and the step height of the DSC curve at the glass transition is relatively small. There is practically no cold crystallization. The quench-cooled samples are amorphous and exhibit a clearer glass transitions and cold crystallization.

In the filled material, all the thermal effects measured are smaller because they relate only to the polymer content. The polymer content can be calculated by comparing the heat of fusion of the filled material with that of the unfilled material according to the equation:

$$\alpha_{polymer} \approx \frac{\Delta H_{fus,filled}}{\Delta H_{fus,unfilled}} \cdot 100\%$$

Here, the polymer content of the filled material is estimated to be 22%.

Conclusion Analysis of the DSC curves provides information on the thermal history, crystallinity and polymer content.

高温聚合物 High Temperature Polymers

样品 聚醚醚酮 PEEK Victrex 450G,注塑成型,
聚醚砜 PES Victrex 200P,注塑成型,
聚四氟乙烯 PTFE 膜,由德国塑料加工协会提供的样品包。

Sample Polyetheretherketone, PEEK Victrex 450G, injection-molded,
Polyethersulfone, PES Victrex 200P, injection-molded,
Polytetrafluoroethylene, PTFE film supplied in the sample set of the "Arbeitsgemeinschaft Deutsche Kunststoff-Industrie"

条件 仪器:DSC **Measuring cell**:DSC
Conditions 坩埚:标准 40μl 铝坩埚,盖钻孔   **Pan**:Aluminum standard 40μl, pierced lid
样品制备: **Sample preparation**:
从注射成型部件上切割出的或从薄膜上冲出的圆片   Disks cut from injection-molded parts or punched from film
DSC 测试: **DSC measurement**:
以 20 K/min 从 30℃加热到 400℃   Heating from 30℃ to 400℃ at 20 K/min
气氛:氮气,50 cm³/min   **Atmosphere**:Nitrogen, 50 cm³/min

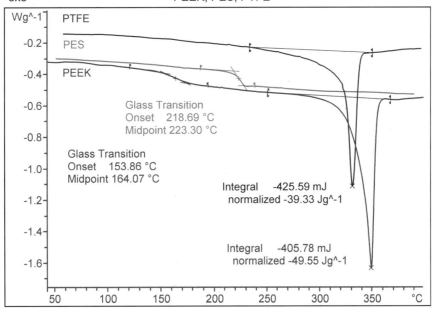

计算
Evaluation

	PEEK	PE	PTFE
T_g,中点,℃ / T_g, midpoint in ℃	164.1	223.3	—
熔融热,J/g / Heat of fusion in J/g	49.6	—	39.3
峰温,℃ / Peak temperature in ℃	349.0	—	330.0
样品质量,mg / Sample mass in mg	8.19	16.39	10.82

解释 PES 是完全无定形的,仅仅显示玻璃化转变(是所有有机材料中最高的之一)。在室温时,PEEK 和 PTFE 是半结晶的,在自动进样器上不能通过骤冷从熔体冻结至玻璃态(当自动进样器将热坩锅放到冷转盘上时平均冷却速率约为 3000K/min)。

DSC 曲线以 W/g 显示。用自动空白扣除,曲线与样品的比热成正比。比例常数即是加热速率 0.333K/s。PTFE 在 60℃的比热约为 0.5J/gK,而 PES 和 PEEK 在 60℃的值约为 1 J/gK。

结论 DSC 的高灵敏度(低信号噪声水平)能容易的测试弱效应。这意味着在研究实验室中新合成的聚合物可用 DSC 快速表征其热行为。

Interpretation PES is completely amorphous and shows only the glass transition (one of the highest of all organic materials). At room temperature, PEEK and PTFE are semicrystalline and cannot be frozen in the glassy state by quench-cooling from melt in the sample robot (the mean cooling rate is about 3000 K/min when the robot places the hot pan on the cold turntable).

The DSC curves are presented in W/g. With automatic blank subtraction this is directly proportional to the specific heat capacity of the samples. The proportionality constant is the heating rate of 0.333 K/s. The specific heat capacity of PTFE is approx. 0.5 J/gK at 60 ℃, whereas the values for PES and PEEK are about 1 J/gK.

Conclusions The high sensitivity (low signal noise level) of the DSC821[e] allows weak effects to be easily measured. This means that new polymers synthesized in research laboratories can be rapidly characterized with respect to their thermal behavior by DSC.

用 DSC 和 TMA 测定聚四氟乙烯多晶态　PTFE Polymorphism by DSC and TMA

样品 **Sample**	直径 4.2mm 的商品 PTFE 棒材 Commercial PTFE rod of 4.2-mm diameter	
条件 **Conditions**	仪器：DSC TMA，球点探头	Measuring cell：DSC TMA with ball-point probe
	坩埚： DSC：40μl 标准铝坩锅，钻孔盖 TMA：样品放置于直径 6 mm 和厚 0.5 mm 的石英片之间	Pan： DSC：aluminum standard 40μl, pierced lid TMA：the sample is mounted between fused silica disks of 6 mm diameter and 0.5 mm thickness
	样品制备： DSC：从棒材上切出的圆片，厚约 0.6mm TMA：高 3.35 mm 的柱状样品作轴向测试。对径向测试，对等两边切掉两部分得到厚 2.42 mm 的中部样品。	Sample preparation： DSC：Disk cut from rod, approx. 0.6 mm thick TMA：cylindrical sample 3.35 mm high for axial measurement. For the radial measurement two parts were cut off from opposite sides so that a 2.42 mm thick center sample was obtained.
	测试： 先加热到 50℃，样品适应石英圆片。在自由冷却至开始温度后进行实际测试，即以 5K/min 从 −20℃（TMA −30℃）加热到 50℃。	Measurements： In the first heating run to 50 ℃, the sample adapts to the silica disks. The actual measurement is performed after uncontrolled cooling to the start temperature, i.e. heating from −20 ℃ (TMA −30 ℃) to 50 ℃ at 5 K/min
	负载：0.05N(TMA) 气氛：氮气，50 cm^3/min（DSC） 氦气，200 cm^3/min（TMA）	Load：0.05 N (TMA) Atmosphere：Nitrogen, 50 cm^3/min (DSC) Helium, 200 cm^3/min (TMA)

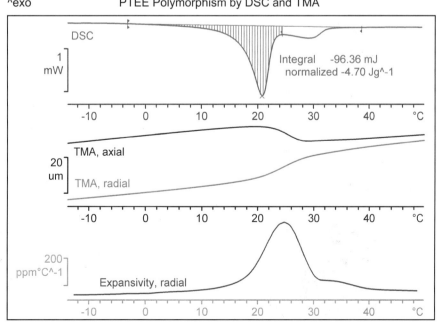

计算 如 TMA 曲线和表格所示,径向膨胀系数显示在转变时最大。在轴向,有对应的最小值(负膨胀系数)。

Evaluation As the TMA curve and table show, the radial expansivity exhibits a maximum during the transition. In the axial direction, there is a corresponding minimum (negative expansivity).

温度,℃ Temperature in ℃	膨胀系数,径向,ppm/K Expansivity, Radial in ppm/K
10	151.6
15	189.2
20	362.9
25	645.6
30	265.7

解释 PTFE 微晶在低于 19℃ 时以三斜晶变体 II 存在。在 19℃ 到 30℃ 之间,六方晶型 IV 是稳定的,加热时,该晶型将转变为六方晶型 I。

制造加工造成 PTFE 棒材是各向异性的,这从沿两个主轴测试的膨胀曲线看是显而易见的。

在 TMA 曲线上,仅有一个 30℃ 处轮廓不清晰的肩。在膨胀系数曲线上 IV-I 转变是清晰可见的。由于热惯性,该曲线比 DSC 曲线宽。

Interpretation The crystallites of PTFE exist below about 19℃ as the triclinic modification II. Between 19 and 30 ℃, the hexagonal form IV is stable. On heating, this transforms to the hexagonal form I.

The rods are anisotropic as a result of the manufacturing process. This is clearly apparent from the dilatometric curves measured along the two main axes.

On the TMA curves, there is only an indistinct shoulder at 30 ℃. The IV-I transition is clearly visible on the expansivity curve. Due to thermal inertia, the curve is broader than the DSC curve.

结论 PTFE 是唯一的的固-固互变时伴随着热焓和体积对应变化的商品聚合物。因此,DSC 和 TMA 可以用来鉴定 PTFE。

Conclusion PTFE is the only commercial polymer with an enantiotropic solid-solid transformation with corresponding enthalpy and volume change. DSC as well as TMA curves can therefore be used to identify PTFE.

用 DMA 和 DSC 表征聚四氟乙烯 Characterization of PTFE by DMA and DSC

样品 聚四氟乙烯 PTFE：a) 白色棒，厚 2.16mm；b) 薄膜，厚 0.2mm
实验目的是用动态热机械分析（DMA）和差示扫描量热法（DSC）研究 PTFE 从 －120℃ 到熔融温度范围内的相转变。

Sample Polytetrafluorethylene, PTFE：a) white rod, 2.16-mm thick；b) film 0.2-mm thick. The purpose of the experiment was to investigate the transitions of PTFE in the temperature range from －120℃ to the melting temperature using both dynamic mechanical analysis (DMA) and differential scanning calorimetry, DSC.

条件 仪器：DMA 和 DSC

Conditions **Measuring cell**：DMA and DSC

样品制备：

Sample preparation：

DMA 剪切形变：从棒材上切出两片直径为6 mm的 2.16 mm 厚的圆片。
DMA 拉伸形变：从膜上切下宽 6.65 mm 的条，夹在样品支架上，提供有效测试长度 10.41 mm。

DMA shear deformation：two disks, 2.16 mm thick of 6 mm diameter, were cut from the rod.
DMA tensile deformation：a 6.65 mm wide strip was cut from the film and clamped in the sample holder to provide an active measurement length of 10.41 mm.

DMA 测试：

剪切测试在 1Hz 和 10Hz 下进行，以 2K/min 的加热速率从 －140℃ 到 300℃。最大力振幅为 8N；最大位移振幅为 10μm；偏移控制为零。
拉伸测试在 1Hz 下进行，以 2K/min 的加热速率从 －140℃ 到 300℃。最大力振幅为 5N；最大位移振幅为 10μm；偏移控制为零。

DMA measurements： The shear measurement was performed at 1 and 10 Hz from －140 ℃ to 300 ℃ at a heating rate of 2 K/min Maximum force amplitude 8 N; maximum displace-ment amplitude 10 μm; offset control zero.

The tensile measurement was performed at 1 Hz and a heating rate of 2 K/min from －140 ℃ to 300℃. Maximum force amplitude 5 N; maximum displacement amplitude 10 μm; offset control zero.

DSC 测试：

在 40μl 铝坩锅中的 25.33 mg PTFE（样品 b）以 10K/min 从 －150℃ 加热到 400℃。

DSC measurement：

25.33 mg of the PTFE (sample b) in a 40 μl aluminum pan were heated from －150℃ to 400 ℃ at 10 K/min.

气氛：

DMA，静态空气；DSC，氮气，50 ml/min。

Atmosphere：

DMA, static air; DSC nitrogen, 50 ml/min.

DMA：tanδ 的峰温。

DMA：peak temperatures of tanδ.

解释 DMA 拉伸测试，在杨氏模量曲线上显示三个主台阶和对应的 tan 曲线上三个主峰。在 DSC 曲线上可以清晰看到室温下转变峰和熔融峰。

Interpretation The DMA tensile measurement shows three main steps in the Young's modulus (E′) curve and three main peaks in the corresponding tan curve. The transition peaks at room temperature and the melting peak can be clearly seen in the DSC curve.

这些转变可以根据 McCrum 等人来鉴别（见本文末参考文献）：

The transitions can be identified as follows according to McCrum et. al. (see reference at the end of the text)：

- －97℃ 处的 γ-松弛
- 19℃ 和 30℃ 处的 β-松弛，它们代

- γ-relaxation at－97 ℃,
- sharp β-transitions at 19 ℃ and 30 ℃, which represent the triclinic

125

表在这两个温度间含有一个无序中间结构的晶体结构中三斜晶型向六方晶型的转变。
- 127℃处的 α-松弛（无定形部分的玻璃化转变）。
- 327℃处的微晶熔融。

to hexagonal change in crystal structure with a disordered intermediate structure between these temperatures,
- α-relaxation (glass transition of the amorphous parts) at 127 °C,
- crystallite melting point at 327 °C.

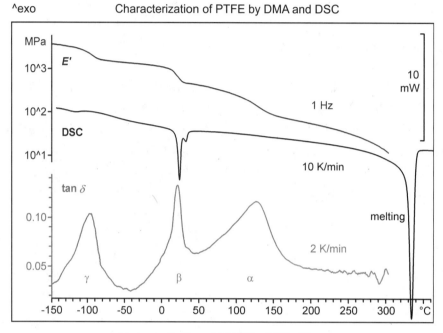

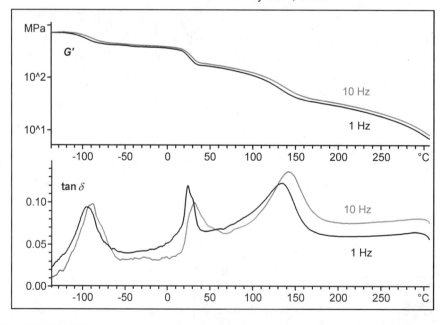

尽管加热速率快，DSC 测试还是分离出了两个尖锐的 β-转变。大的吸热峰显示熔融。然而，γ-转变并没有清晰的看到。在给定条件下由于

The DSC measurement separates the two sharp β-transitions despite the fast heating rate. The large endothermic peak indicates the melting. However, the γ-transition is not clearly seen and the glass transition is too small to be detected by heat

玻璃化转变太小而不能被热流测试检测到。

DMA测试显示熔融温度以下的所有的三个转变。在拉伸测试中β-转变没有被分开。在剪切测试中，两个转变显示为在相应1Hz和10Hz的tan峰中的两个肩(见第二张图)。

剪切模量(G′)和各个tanδ曲线显示了通常的频率依赖性，如从第二张图中能见到的。如同由泊松比所预期的，杨氏模量E′总是比G′大。

用DSC和DMA测试的转变温度总结在下表中，并与所引用的参考文献中的给定值进行比较。

flow measurements under the given conditions.

The DMA measurements show all three transitions below the melting temperature. The β-transition is not resolved in the tensile measurement. In the shear measurement, the two transitions are shown as shoulders in the corresponding tanδ peaks at 1 and 10 Hz (see the second diagram).

The shear modulus (G′) and the respective tanδ curves show the usual frequency dependence as can be seen in the second diagram. Young's modulus, E′, is always greater than G′, as expected from Poisson's ratio.

The transition temperatures measured by DSC and DMA are summarized in the following table and are compared with the values given in the reference cited.

	转变 Transition		γ	β1	β2	α	熔融 melting
DSC	峰 peak	℃	~−100	24.1	32.5		
DSC	起始点 onset	℃					328
DMA 拉伸 DMA tensile	1Hz 时 tanδ tanδ at 1 Hz	℃	−96	22		126	
DMA 剪切 DMA shear	1Hz 时 tanδ tanδ at 1 Hz	℃	−96	24		133	
DMA 剪切 DMA shear	10Hz 时 tanδ tanδ at 10 Hz	℃	−89		32	141	
参考文献（剪切测试）Reference (shear meas.)	1Hz 时 tanδ tanδ at 1 Hz	℃	−97	19	30	127	327

结论 用DMA的剪切模式，能在宽刚度范围内测量模量G′和tanδ。PTFE的转变可以很容易地表征，即使在非常低的温度下。剪切或拉伸形变都可以使用，两者给出相同的结果。DMA测试能显示用DSC观察不到的转变。DMA具有研究频率对转变影响的优势。

参考文献 N. G. McCrum, B. E. Read, G. Williams：Anelastic and Dielectric Effects in Polymeric Solids, Dover Publications, Inc., New York (1991)，450-460页

Conclusions Using the DMA/SDTA861e in the shear mode, the modulus, G′, and tanδ can be measured over a wide stiffness range. Transitions in PTFE can be easily characterized, even at very low temperatures. Either shear or tensile deformation can be used; both give the same results. The DMA measurements show transitions not observed by DSC. DMA has the advantage that the influence of frequency on the transitions can be studied.

Reference N. G. McCrum, B. E. Read, G. Williams, Anelastic and Dielectric Effects in Polymeric Solids, Dover Publications, Inc., New York (1991) pages 450-460.

用 ADSC 测定聚醚酰亚胺的玻璃化转变　PEI, Glass Transition by ADSC

样品	聚醚酰亚胺 PEI Ultem 1000 注射-成型部件
Sample	Polyetherimide, PEI, Ultem 1000, injection-molded part
条件	仪器：DSC
Conditions	**Measuring cell**: DSC with sample changer
	坩锅：40 μl 标准铝坩锅
	Pan: Aluminum standard 40 μl
	样品制备：
	Sample preparation:
	用刀切成底部是平的,17.546 mg
	17.546 mg cut with knife so that the bottom is flat
	DSC 测试：
	DSC measurements:
	以平均速率 1K/min 从 200℃ 加热到 229℃。正弦振动振幅 1K,周期 1 min。
	Heating from 200℃ to 229℃ at a mean rate of 1 K/min. Amplitude of sinusoidal oscillation 1 K, period 1 min
	气氛：氮气,50 cm³/min
	Atmosphere: Nitrogen, 50 cm³/min

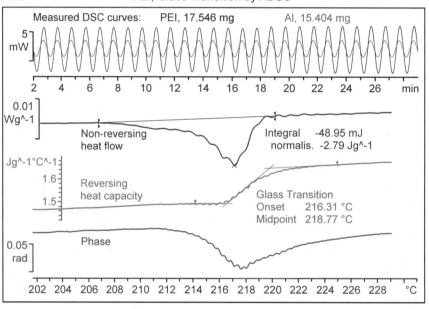

计算　ADSC 的计算基于快速傅立叶分析(FFT)。除了所显示的曲线外,还可显示比热的实数和虚数部分的曲线。
每一条计算得到的曲线都可以进行数学计算,例如：积分不可逆峰得到松弛峰的焓变。玻璃化转变是从 c_p 温度函数计算的,c_p 值示于表格中：

Evaluation　The evaluation of the ADSC curves is based on fast Fourier analysis (FFT). In addition to the curves displayed, curves of the real and imaginary components of the specific heat capacity could also be shown.
Each calculated curve can be numerically evaluated, e.g. the non-reversing peak integrated to get the enthalpy change of the relaxation peak. The glass transition is evaluated from the C_p temperature function and C_p is shown in tabular form:

温度,℃ Temperature in ℃	C_p, J/gK
210	1.48
215	1.49
220	1.62
225	1.66

解释 最上面的曲线为正弦的 ADSC 曲线：红色曲线是用 15.404mg 铝作为热容参比得到的，黑色曲线是用加入了 17.546mg PEI 得到的。两条曲线都已经过空白修正。蓝色曲线显示焓松弛的"不可逆"热流峰。比热曲线（绿色）是"可逆"曲线。样品对于铝参比样的相角偏移显示在棕色曲线中。

结论 如本例所示，ADSC 分离所谓的"可逆"和"不可逆"效应。这对有重叠的反应可能是非常有用的。通常，C_p 变化为可逆效应，例如在玻璃化转变处。不可逆效应是焓松弛、冷结晶和所有种类的化学反应。

Interpretation The uppermost curves display the sinusoidal ADSC curves; the red one is obtained from the 15.404 mg aluminum used as a heat capacity reference, and the black one from the additional 17.546 mg PEI. Both curves have been blank-corrected. The blue curve shows the "non-reversing" heat flow peak of the enthalpy relaxation. The specific heat capacity curve (green) is the "reversing" curve. The phase shift of the sample with respect to the aluminum reference sample is shown in the brown curve.

Conclusions As the example demonstrates, ADSC separates so-called "reversing" and "non-reversing" effects. This can be very useful with reactions that overlap. In general, cp changes are reversing effects, e.g. at the glass transition. Non-reversing effects are enthalpy relaxation, cold crystallization and all kinds of chemical reactions.

聚醚酰亚胺的 DMA 分析　DMA Analysis of PEI

样品　两个聚醚酰亚胺 PEI 薄膜，用动态热机械分析 DMA 进行比较。样品名为 A 和 B。

Sample　Two thin films of polyetherimide, PEI, for comparison by dynamic mechanical analysis, DMA. The samples were named A and B.

条件 Conditions	仪器： DMA，拉伸样品支架	Measuring cell： DMA with the tension sample holder
	样品制备： 薄膜，长 10.5mm，厚 22μm，宽 5.0 和 3.6 mm（分别为样品 A 和 B）	Sample preparation： Thin films, length 10.5 mm, thickness 22μm, width 5.0 and 3.6 mm (samples A and B, respectively)
	DMA 测试： 分别以 4 K/min（样品 A）和 6 K/min（样品 B）从 -150℃ 到 500℃，用 1Hz 频率。最大力振幅 1N；最大位移振幅 25μm；偏移控制为零。	DSC measurements： -150℃ to 500 ℃ at 4 K/min (sample A) and 6 K/min (sample B) respectively, using a frequency of 1 Hz. Maximum force amplitude 1 N; maximum displacement amplitude 25 μm; offset control zero.
	气氛：氮气，200 cm^3/min	Atmosphere：Nitrogen, 200 cm^3/min

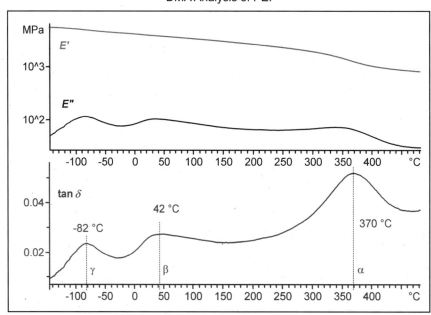

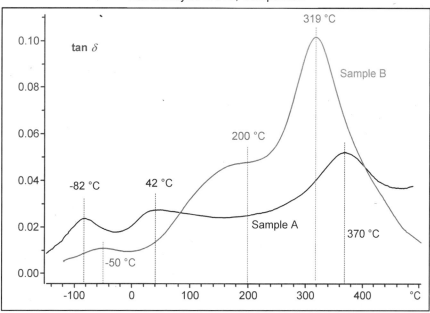

DMA Analysis of PEI, Comparison

计算 tanδ 的峰温和 E' 的绝对值
Eualuation Peak temperatures of tan δ and absolute value of E'

	样品 A Sample A	样品 B Sample B
α-松弛（玻璃化转变），℃ α-relaxation (glass transition) in ℃	370	319
β-松弛，℃ β-relaxation in ℃	42	200
γ-松弛，℃ γ-relaxation in ℃	−82	−50
−100℃时的杨氏模量 E'，GPa Young's modulus, E', at −100 ℃ in GPa	5.09	9.93

解释 两个样品都显示三个松弛区域。样品 B 的玻璃化转变发生在 319℃，样品 A 的在 370℃。玻璃化转变以下，观察两个更早的松弛现象。样品 A 和样品 B 松弛间隔的宽度即α-松弛和β-松弛之间的温差是明显不同的，这表明两种样品在内部结构方面有相当大的不同。

在 −100℃，样品 B 的模量的绝对值是样品 A 的两倍（见上面的表格）。

结论 材料的松弛现象可用 DMA 来研究。这类实验可以得出有关材料的分子结构方面的结论。本实验结果清楚的显示了两种 PEI 膜的不同行为。两个材料都具有非常高的玻璃化转变温度。

Interpretation Both samples show three relaxation regions. The glass transition of sample B occurs at 319℃ and that of sample A at 370℃. Below the glass transition, two further relaxation phenomena are observed. The width of the intermediate relaxation, i.e. the temperature difference between the-and the-relaxation differs quite significantly from sample A to sample B. This indicates that the two samples differ considerably in their structure.

At −100 ℃, the absolute value of the modulus of sample B is twice that of sample A (see table above).

Conclusions Relaxation phenomena in materials can be investigated by DMA. Experiments of this type allow conclusions to be drawn about the molecular structure of the material. The results of this experiment show clear differences in the behavior of the two PEI films. Both have a very high glass transition temperature.

7.8 热塑性弹体 Thermoplastic Elastomers

酯类热塑性弹体 TPE-E, Ester-based Thermoplastic Elastomer

样品 Hytrel HTR 8105 BK 球状颗粒

$$\left[-\left[\begin{array}{c}R\\|\\O-CH-(CH_2)_n\end{array}\right]_k -O-\overset{O}{\underset{}{C}}-\bigcirc-\overset{O}{\underset{}{C}}-\right]_n -\left[-O-(CH_2)_4-O-\overset{O}{\underset{}{C}}-\bigcirc-\overset{O}{\underset{}{C}}-\right]_m$$

柔软链段 Soft Segment   坚硬链段 Rigid Segment

Sample Hytrel HTR 8105 BK as pellets

条件 / Conditions

测试仪器：DSC	**Measuring cell**: DSC
坩锅：40μl 标准铝坩锅，钻孔盖	**Pan**: Aluminum standard 40μl, pierced lid
样品制备：	**Sample preparation**:
从球状颗粒切出的圆片，7.3mg	Disk cut from pellet, 7.3 mg
DSC 测试：	**DSC measurements**:
以 20K/min 从 30℃加热到 250℃	Heating from 30℃ to 250 ℃ at 20 K/min
气氛：氮气，50 cm³/min	**Atmosphere**: Nitrogen, 50 cm³/min

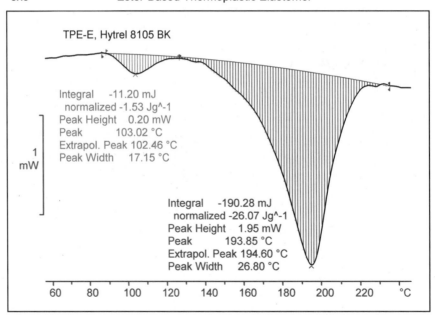

计算 / Evaluation

	第一个峰 First peak	第二个峰 Second peak
熔融热, J/g Heat of fusion in J/g	1.5	26.1
峰温, ℃ Peak temperature in ℃	103.0	193.8

使用样条基线，它与峰前和峰后的 c_p 温度函数吻合很好。

The spline baseline type was used. It fits the c_p temperature function before and after the peaks very well.

解释 DSC 曲线在测试的温度范围内显示两个吸热峰。峰与大分子的坚硬链段的熔融相关联。

结论 TPE 给出能用来鉴别的特征 DSC 曲线。要表征软段的弹性行为应使用低温 DSC(−100 ℃)。

Interpretation The DSC curve shows two endothermic peaks in the temperature range measured. The peaks are related to the melting of the rigid segments of the macromolecule.

Conclusions TPE gives characteristic DSC curves that can be used for identification purposes. Low temperature DSC (−100 ℃) would have to be used to characterize the elastomeric behavior of the soft segments.

烯烃类热塑性弹体 TPE-O, Olefin-based Thermoplastic Elastomer

样品　　PP/EPDM 共混物：
　　　　　　Santoprene 101-55,
　　　　　　Santoprene 101-80 和
　　　　　　Santoprene 103-40

$$\left[-(CH_3)_k-\underset{\underset{CH_3}{|}}{CH}-CH_2)_n-(CH_2-\underset{\underset{\underset{CH_3-CH=CH}{|}}{CH_2}}{CH})_m-\right]_p \left[-\underset{\underset{CH_3}{|}}{CH}-CH_2-\right]_n$$

　　　　　　柔软链段 soft segment　　　　　坚硬链段 Rigid segment

Sample　PP/EPDM blends:
　　　　　　Santoprene 101-55,
　　　　　　Santoprene 101-80 and
　　　　　　Santoprene 103-40

条件　　仪器：DSC　　　　　　　　　　　**Measuring cell**: DSC
Conditions　坩锅：40 μl 标准铝坩锅，钻孔盖　　**Pan**: Aluminum standard 40 μl, pierced lid
　　　　　　样品制备：从球状颗粒切出圆片　　**Sample preparation**: Disk cut from pellet
　　　　　　DSC 测试：　　　　　　　　　　　**DSC measurements**:
　　　　　　以 10 K/min 从 30℃ 加热到 150℃　Heating from 30℃ to 150℃ at 10 K/min
　　　　　　气氛：氮气，50 cm³/min　　　　　**Atmosphere**: Nitrogen, 50 cm³/min

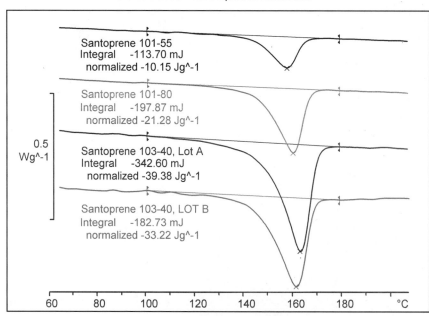

计算
Evaluation

Santoprene	101-55	101-80	103-40,A 批	103-40,B 批
熔融热,J/g Heat of fusion in J/g	10.2	21.3	39.4	33.2
峰温,℃ Peak temperature in ℃	157.2	159.6	162.4	161.1
样品质量,mg Sample mass in mg	11.2	9.3	8.7	5.5

解释 在测试的温度范围内唯一明显的热效应为聚丙烯部分的熔融。

结论 聚丙烯部分的熔融热是所研究的共混物的特征,对质量控制有用。A 批和 B 批的加工性能是不同的。标准的质量是 A 批。B 批的熔融热较低,在相同的吹塑成型加工条件下生产的产品壁较薄。出于安全的原因,B 批被拒绝并退给供应商。

Interpretation The only significant thermal effect in the temperature range measured is the melting of the polypropylene fraction.

Conclusions The heats of fusion of the polypropylene fractions are characteristic for the blends investigated and are useful for quality control purposes. The processability of Lots A and B is different. The standard quality is Lot A. Lot B has a lower heat of fusion and yields a product whose wall thickness is thinner under identical blow-molding processing conditions. For safety reasons, Lot B was rejected and returned to the supplier.

7.9 聚合物共混物和共聚物 Polymer Blends and Copolymers

用 DSC 测试聚碳酸酯和聚碳酸酯/ABS 共混物
PC and a PC/ABS Blend Measured by DSC

样品 PC Makrolon 2405；
PC/ABS 共混物 Bayblend T85

Sample PC，Makrolon 2405；
PC/ABS Blend Bayblend T85

条件
测试仪器：DSC
坩埚：40μl 标准铝坩埚，钻孔盖
样品制备：从球状颗粒切下圆片
DSC 测试：
第一轮以 20K/min 从 30℃ 到 250℃ 以消除热历史，以 20K/min 冷却
实际测试：
以 20K/min 从 30℃ 加热到 250℃。
气氛：氮气，50 cm³/min

Conditions
Measuring cell：DSC
Pan：Aluminum standard 40μl, pierced lid
Sample preparation：Disk cut from pellet
DSC measurements：
First run 30℃ to 250℃ at 20 K/min to eliminate the thermal history，cooling at 20 K/min
Actual measurement：heating from 30℃ to 250 ℃ at 20 K/min
Atmosphere：Nitrogen，50 cm³/min

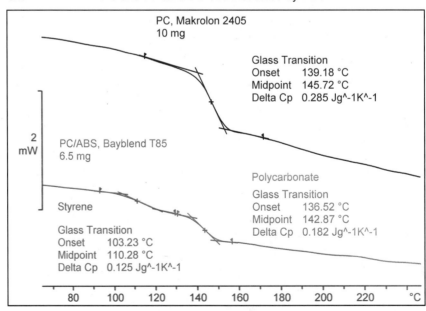

计算 表中用50%转变时的中点给出玻璃化转变温度。
Evaluation In the table the glass transition temperatures are given as midpoints at 50% transition.

	Makrolon2405	Bayblend T85
T_g, PC, ℃	145.7	142.9
Δcp, PC, J/gK	0.29	0.18
T_g, ABS, ℃	—	110.3
样品质量, mg Sample mass in mg	10.0	6.5

从PC玻璃化转变的台阶高度的比例（$\Delta C_{ppure}/\Delta C_{pmixture}$），可以估算混合物由三分之二的PC和三分之一的ABS组成。

From the ratio of the step heights of the PC glass transitions ($\Delta C_{ppure}/\Delta C_{pmixture}$), it can be estimated that the mixture consists of two thirds PC and on third ABS.

解释 PC是一种无定形聚合物，在150 ℃左右有明显的玻璃化转变。在共混物中，加入的聚合物的特征效应是明显的。在ABS例子中，主要是聚苯乙烯的玻璃化转变。Bayblend T85是两种组分都以分离相存在于其中的不相容（不混溶）的聚合物，它们显示各自的玻璃化转变。然而，PC的玻璃化转变温度由于与ABS的相互作用降低了约3 K。

Interpretation PC is an amorphous polymer with a distinct glass transition around 150 ℃. In blends, the characteristic effects of the added polymers become noticeable. In the case of ABS it is mainly the glass transition of polystyrene. Bayblend T85 is an incompatible (immiscible) polymer in which both components are present as separate phases. These exhibit individual glass transitions. The PC glass transition temperature is however lowered by about 3 K due to interaction with the ABS.

结论 聚碳酸酯和它的一些共混物可以容易地用DSC进行测试。某些PC膜（例如：Makrofol）是半晶态的，在235 ℃左右显示一个小的熔融峰。

Conclusions Polycarbonate and some of its blends can easily be measured by DSC. Certain PC films (e.g. Makrofol) are semicrystalline and exhibit a small fusion peak around 235 ℃.

用 DSC 和 TMA 表征乙烯/醋酸乙烯共聚物
E/VAC, Characterization by DSC and TMA

样品 **Sample**	条状乙烯-醋酸乙烯酯共聚物，厚约 1.7 mm Ethylene-vinylacetate copolymer in the form of strips approx. 1.7-mm thick	
条件 **Conditions**	测试仪器：DSC 和 TMA	**Measuring cell**：DSC and TMA
	坩埚： DSC 40 μl 标准铝坩锅，TMA 中是样品上面下面放石英圆片	Pan： Aluminum standard 40μl for DSC, fused silica disks above and below the sample in the TMA
	样品制备： DSC：冲出一个直径 6 mm 的圆片，20.117mg TMA：冲出一个直径 6 mm 的圆片，放在石英片之间。3 mm 的圆点探头放在夹层结构上	Sample preparation： DSC：punching a disk of 6 mm diameter, 20.117 mg TMA：punching a disk of 6 mm diameter that is placed between fused silica disks. Probe with 3 mm ball-point on the sandwich
	测试： 第一轮和第二轮都以 10 K/min 从 −80℃ 加热到 120℃，不控制冷却到 −80℃	Measurement： First and second heating runs from −80℃ to 120℃ at 10 K/min Uncontrolled cooling to −80℃
	负载：0.02 N	Load：0.02 N
	气氛：DSC：氮气，50 cm³/min	Atmosphere：DSC：Nitrogen, 50 cm³/min

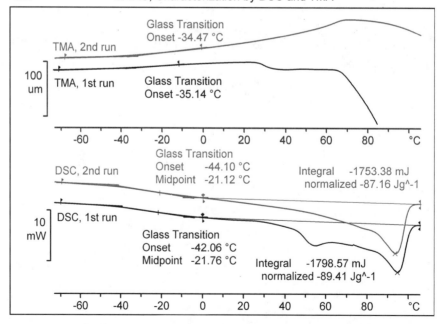

计算
Evaluation

玻璃化转变温度表:
Table of glass transition temperatures:

	DSC 起始点,℃ DSC Onset in ℃	DSC 中点,℃ DSC Midpoint in ℃	TMA 切线交点,℃ TMA Intersection in ℃
第一轮 1st run	−42.1	−21.8	−35.1
第二轮 2nd run	−44.1	−21.1	−34.5

DSC 曲线的熔融峰对应于 PE 的结晶区域。用直线（基线）从 0℃ 到 110℃ 进行积分。虽然形状不同，但熔融热大致相同，分别为 89 J/g 和 87 J/g。将这个值除以对低峰温 PE 假定的熔融热 100 J/g 给出 PE 的含量为 82%。

The DSC fusion peaks correspond to the crystalline regions of the PE. Integration is performed from 0℃ to 110℃ using a straight baseline. Despite the different shapes, the heat of fusion is about the same, namely 89 J/g and 87 J/g respectively. Dividing this value by an assumed heat of fusion (100 J/g) for PE with such a low peak temperature gives a PE content of 82%.

第二轮 TMA 得到下列膨胀系数结果:

The second TMA run yielded the following results for expansivity:

温度,℃	−50	−25	0	25	50
膨胀系数,ppm/K	94	207	276	364	559

Temperature in ℃	−50	−25	0	25	50
Expansivity in ppm/K	94	207	276	364	559

解释 在 −40℃ 左右的玻璃化转变，第一轮和第二轮是相似的。DSC 熔融峰的形状是不同的。用贮存于 25℃ 的样品测试的第一轮在 55℃ 显示熔融中断（参见应用范例"聚乙烯的熔融曲线和热历史"）。在贮存期间形成的结构在约 55℃ 时熔融，产生该额外的峰。

在玻璃化转变以上，TMA 曲线是完全不同的。在第一轮加热期间，软薄膜的蠕变减小了石英圆片之间的距离。在 65℃，低压缩应力 0.02N/28mm² 即 0.7 mN/mm² （直径 6mm）下的塑性变形开始。第二轮显示了明显的膨胀，随后在 65℃ 尺寸稳定性丧失。

Interpretation Around the glass transition at −40℃, the first and the second runs are similar. The DSC fusion peaks are different in shape. The first run with the sample stored at 25℃ shows a melting gap at 25℃ (see Application Example "PE, melting curve and thermal history"). The structures formed during the storage period melt at about 55℃ and cause the additional peak.

Above the glass transition, the TMA curves are quite different. During the first heating run the creep of the soft film reduces the distance between the fused silica disks. At 65℃, plastic deformation begins under the low compression stress of 0.02 N/28 mm², i.e. 0.7 mN/mm² (diameter 6 mm). The second run exhibits a pronounced expansion followed by the loss of dimensional stability at 65℃.

结论 DSC 和 TMA 的玻璃化转变温度是不同的，因为两个技术测试的是不同的性能。DSC 熔融热可以估算 PE 的含量。用已知共聚物校准后，更精确的测定是可能的。
TMA 曲线还得到膨胀系数和尺寸稳定性丧失的"热变形温度"。

Conclusions The DSC and TMA glass transition temperatures are different because the techniques measure different properties. The DSC heat of fusion allows the PE content to be estimated. More accurate determinations would be possible after calibration with known copolymers.
The TMA curves also yield the expansivity and a "heat distortion temperature" at which dimensional stability disappears.

用 DSC 测定丙烯腈/丁二烯/苯乙烯共聚物的玻璃化转变
ABS Glass Transition by DSC

样品　Terluran 996S，Cycolac TCA Q，Lego"砖"，Novodur P2M 和 Magnum 3416 HH
Samples　Terluran 996S，Cycolac TCA Q，Lego "brick"，Novodur P2M and Magnum 3416 HH

条件　测试仪器：DSC　　　　　　　　　　**Measuring cell**：DSC
Conditions　坩埚：40 μl 标准铝坩埚，钻孔盖　　**Pan**：Aluminum standard 40 μl, pierced lid

样品制备：
从球状颗粒或注射成型部件上切下圆片

Sample preparation：
Disk cut from pellet or injection-molded part

DSC 测试：
在第一轮加热中，通过从 30℃ 加热到 250℃ 消除热历史。不受控冷却到室温后，以 20K/min 和 10K/min 进行实际测试，后者的起始温度为 −150℃。

DSC measurement：In a first heating run, the thermal history is eliminated by heating from 30℃ to 250℃. After uncontrolled cooling to room temperature, the actual measurement is performed at 20 K/min and at 10 K/min. With the latter, the initial temperature was −150℃.

气氛：氮气，50 cm³/min
Atmosphere：Nitrogen，50 cm³/min

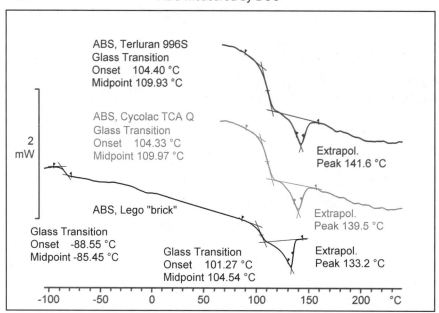

计算
Evaluation

名称 Name	样品质量 mg Sample mass in mg	丁二烯 T_g ℃ Tg butadiene in ℃	苯乙烯 T_g ℃ Tg styrene in ℃	丙烯腈峰温 ℃ Acrylonitril peak temperature, in ℃
Terluran 996S	7.6	—	109.9	141.6
Cycolac TCA Q	6.1	—	110.0	139.5
Lego "砖" Lego "brick"	10.9	−85.5	104.5	133.2
Novodur P2M	7.4	—	108.1	—
Magnum 3416 HH	6.1	—	122.9	—

解释 ABS 最明显的热效应是 100℃左右的聚苯乙烯组分的玻璃化转变。在低温区域,DSC 也能检测到聚丁烯的转变。在 100℃以上,有一个聚丙烯腈产生的峰,因为是一个峰而不是 c_p 变化,所以推荐用峰温进行表征。聚丙烯腈含量非常低的 ABS 不会显示这个峰。

注:有些作者将 130℃左右的峰解释为滑爽添加剂或者成型润滑剂的熔融。

Interpretation The most pronounced thermal effect of ABS is the glass transition of the polystyrene component around 100℃. In the low temperature range, DSC can also detect the polybutadiene transition. Above 100℃, there is a peak caused by the polyacrylonitrile. Since it is a peak rather than a change in c_p, the peak temperature is recommended for characterization. ABS with a very low content of polyacrylonitrile does not exhibit this peak.

Note: some authors interpret the peak around 130℃ as being due to melting of a slip additive or mold lubricant.

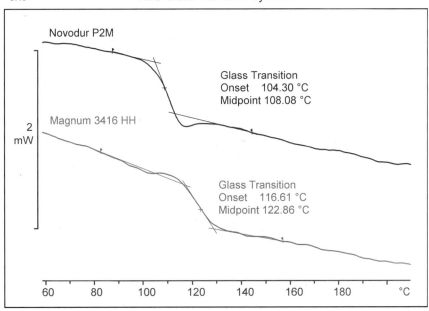

结论 ABS 样品的 DSC 曲线是不同的,可用来表征材料。因此 DSC 是 ABS 来料质量控制的好方法。

Conclusion The DSC curves of the ABS samples are different and can be used to characterize the materials. DSC is therefore a good method for the quality control of incoming ABS material.

甲基丙烯酸甲酯/丁二烯/苯乙烯共聚物的 DSC 和 DMA 测试
DSC and DMA Measurements of an MBS Copolymer

样品　　未知组成的甲基丙烯酸甲酯-丁二烯-苯乙烯(MBS)共聚物粉末。分析的目的是测定玻璃化转变。

Sample　Methylmethacrylate-butadiene-styrene (MBS) copolymer powder of unknown composition. The purpose of the analysis was to determine the glass transitions.

条件　测试仪器:DSC 和 DMA

Conditions　**Measuring cell**:DSC and DMA

样品支架:
DSC:40μl 标准铝坩锅
DMA:剪切样品支架
样品制备:
DSC:12.69 mg 粉末样品放入坩锅内,盖钻孔。
DMA:粉末被压成直径 5.0 mm 的 1.71 mm 厚圆片。两个圆片被装进剪切夹具。
DSC 测试:
以 20K/min 从 −100℃ 加热到 150℃,随后以 5K/min 从 150 冷却到 −100℃。
DMA 测试:
以 2K/min 从 −100℃ 加热到 180℃
最大力振幅 10N;最大位移振幅 1μm;偏移控制零。
测试在频率系列 100Hz、10Hz 和 1Hz 下进行。
气氛:静态空气

Sample holder:
DSC:40μl standard aluminum crucibles
DMA:shear sample holder
Sample preparation:
DSC:12.69 mg of the powder were placed in a crucible with pierced lid.
DMA:The powder was pressed to 1.71 mm thick disks of 5.0 mm diameter. Two disks were mounted in the shear clamp.
DSC measurement:
Heating from −100℃ to 150℃ at 20 K/min, with subsequent cooling from 150 to −100℃ at 5 K/min
DMA measurement:
Heating from −100℃ to 180℃ at 2 K/min
Maximum force amplitude 10N; maximum displacement amplitude 1μm; offset control zero.
The measurement was performed in a frequency series at 100, 10 and 1 Hz.
Atmosphere:Static air

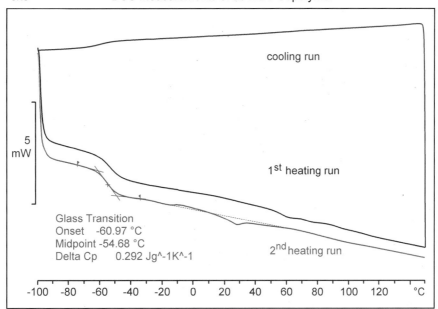

解释　玻璃化转变用 DSC 这样的量热方法有时候是难以测量的。例如，对含有一个低含量组分的共聚物只能检测到一个转变。MBS 的 DSC 测试显示在 -54.8 ℃处聚丁二烯的玻璃化转变。在 -20 ℃聚丁二烯开始结晶，随后熔融。其它共聚物组分的第二个玻璃化转变不能在测试曲线上清晰地看到。

Conditions　Glass transitions are sometimes difficult to measure with calorimetric methods such as DSC. For example, often only one transition can be detected in copolymers that have a low content of one component. The DSC measurement of MBS shows the glass transition of the polybutadiene at -54.8 ℃. Crystallization of the polybutadiene begins at -20 ℃ followed by melting. A second glass transition of the other copolymer components cannot be clearly seen in the measurement curve.

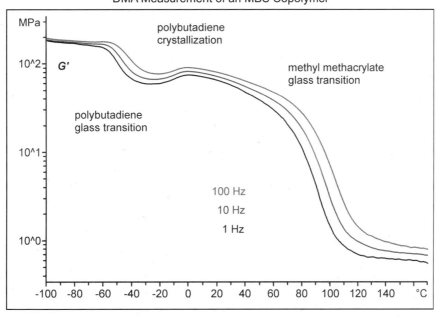

DMA 通常是用于测定这些玻璃化转变的较好的方法，因为在松弛过程中机械性能的变化比量热性能要剧烈得多。在三个频率下用剪切模式的 MBS 的 DMA 测试中，随着储能模量的降低观察到聚丁二烯组分的玻璃化转变。这出现在 DSC 测试中的相同温度。在这个低温玻璃化转变之后，模量由于一个与频率无关的效应而提高，这就是在 DSC 曲线上已经观察到的结晶。晶体在 10℃ 开始熔融，模量轻微下降。在 60℃ 以上，材料由于第二个玻璃化转变而软化，可以清楚地看到储能模量的大幅降低。这个转变也依赖于测试频率，如同在第一个玻璃化转变中观察到的。这可以用由于机械形变所引起的协同重排需要一定的时间来发生的事实作出解释。因此这种所谓的松弛在较高频率下需要较高的运动度。这只能在较高的温度才能得到。所以频率从 1Hz 提高到 100Hz，玻璃化转变温度移动了近 15K。

结论 DMA 是测定玻璃化转变的更好方法，尤其对聚合物。用量热的方法常常难以或者甚至不可能检测到所有的玻璃化转变。在这样的情况下，动态热机械分析是正确的选择。MBS 共聚物的 DMA 曲线仅仅显示两个玻璃化转变。这是由于甲基丙烯酸甲酯和苯乙烯组分的转变温度非常接近而相互重叠，不能用这些测试分开。

DMA is often a better method to use to detect these glass transitions because the mechanical properties change much more dramatically during relaxation processes than calorimetric properties. In the DMA measurement of MBS in shear mode at three different frequencies, the glass transition of the polybutadiene component is observed as a decrease of the storage modulus. This occurs at the same temperature as in the DSC measurement. After this low-temperature glass transition, the modulus increases due to an effect that is independent of frequency. It corresponds to the crystallization already observed on the DSC curve. The crystals begin to melt at 10℃ and the modulus decreases slightly. Above 60℃, the material softens due to the second glass transition, which can clearly be seen as a large decrease in the storage modulus. This transition also depends on the measurement frequency just as is observed with the first glass transition. This can be explained by the fact that the cooperative rearrangements due to the mechanical deformation need a certain amount of time to take place. This so-called relaxation therefore requires a higher degree of mobility at the higher frequency. This is only available at higher temperatures. The increase in frequency from 1 to 100 Hz therefore shifts the glass transition temperature by approximately 15 K.

Conclusions DMA is a better method to detect glass transitions, especially in polymers. Often, it is difficult or even impossible to detect all the glass transitions by calorimetric methods. In such cases, dynamic mechanical analysis is the right choice. The DMA curve of the MBS copolymer shows only two glass transitions. This is due to the fact that the transition temperatures of the methyl methacrylate and the styrene components are very close and overlap each other and cannot be separated by these measurements.

聚ε-己内酰胺/聚四氢呋喃共聚物的结晶和熔融
Crystallization and Melting of PCL/PTHF Copolymers

样品 比例改变的三种不同 PCL/PTHF 共聚物：
A:100% PCL(改性的);B:90:10 PCL/PTHF;C:50:50 PCL/PTHF。
研究的目的是测试样品从熔体冷却时的结晶温度和测定共聚物对模量的影响。
背景：通常骨折修补时用浸过熟石膏(部分脱水的石膏)的纱布绷带包扎固定在适当的位置。但是这种材料的一个严重缺点是它一遇到水会失去刚度。所以寻找医用熟石膏的替代品，开发了一种由聚 ε-己内酰胺(PCL)和聚四氢呋喃(PTHF)组成的合适共聚物。这种材料(PCL/PTHF)可以在体温下使用，具有合适的刚度，与水接触不敏感。

Sample Three different PCL/PTHF copolymers in various ratios:
A: 100% PCL (modified); B: 90:10 PCL/PTHF; C: 50:50 PCL/PTHF.
The purpose of the investigation was to measure the crystallization temperature of the samples when cooled from the melt and to determine the influence of the copolymer on the modulus.
Background: Normally, bone fractures are held in place while they mend by encasing them in gauze bandages dipped in plaster of Paris (partly dehydrated gypsum). One serious disadvantage of this material, however, is that it loses stiffness if it comes in contact with water. Alternatives to plaster of Paris for medical applications were therefore sought. A suitable copolymer was developed consisting of poly-ε-caprolactone (PCL) and polytetrahydrofurane (PTHF). This material (PCL/PTHF) can be applied at body temperature, has the right stiffness and is insensitive to contact with water.

条件 测试仪器:
Conditions DMA,剪切夹具样品支架
DSC,40μl 铝坩锅
样品制备:
切出厚 2.1mm 大约 10×10 mm 的矩形样品，夹入 DMA 的剪切夹具样品支架内。
DMA 和 DSC 温度程序:
以 2K/min 从室温加热到 80℃，然后在 80℃ 恒温保持 10 分钟，最后以 2K/min 从 80℃冷却到-60℃。
DMA 测试:
最大力振幅 40N;最大位移振幅 3μm;频率 1Hz;偏移控制零

气氛:静态空气

Measuring cell:
DMA with shear clamp sample holder
DSC with 40 μl aluminum pans
Sample preparation:
Rectangular samples 2.1 mm thick and about 10×10 mm were cut out and clamped in the shear clamp sample holder of the DMA.
DMA and DSC temperature program:
Heating from room temperature to 80℃ at 2 K/min, then held isothermally at 80℃ for 10 min, and finally cooling from 80 to -60℃ at 2 K/min
DMA measurement:
Maximum force amplitude 40 N; maximum displacement amplitude 3 μm; frequency 1 Hz; offset control zero
Atmosphere: Static air

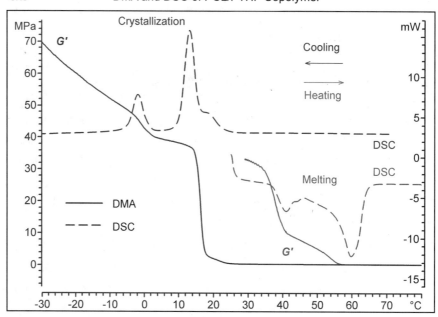

解释 第一张图显示样品 C 的 DMA 加热/冷却实验的剪切模量（G'）和对应的 DSC 热流曲线。加热期间，测试到两个对应于两种组分熔融的吸热效应。模量在第一个热效应期间下降，材料在 40℃ 左右已经软化。

Interpretation The first diagram shows the shear modulus (G') of sample C from the DMA heating/cooling experiment and the corresponding DSC heat flow curves. During heating, two endothermic effects are measured that correspond to the melting of the two components. The modulus drops during the first effect and the material is already soft at about 40℃.

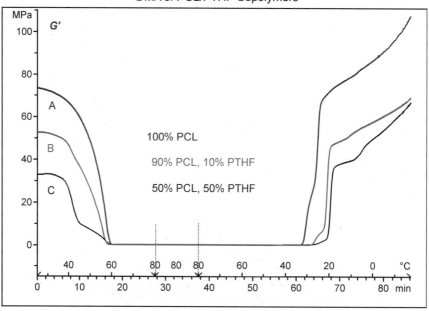

冷却过程中的开始两个峰显示 PCL 结晶分两个阶段；第三个峰显示的是 PTHF 的结晶。在这些转

The first two peaks during cooling show that PCL crystallizes in two stages; the third peak shows the crystallization of PTHF. During these transitions, the material regains the original

变中,按照模量上升显示的,材料重新获得原始刚度。

在第二张图中,比较了三个样品在熔融和结晶时的剪切模量(G')。固态模量随着 PTHF 含量的增加而降低。模量似乎是低的,而实际上是调节到这个值使材料对应用仍然有足够的刚硬,但另一方面不太脆,在受到机械冲击时不会破裂。

PTHF 的水平也决定了结晶的温度和时间。随着 PTHF 含量的提高,结晶向较低温度移动,即等温结晶时间较长。

这些结果与用相同温度范围的时间分辨小角(SAXS)和广角(WAXS)X-光散射分析获得的其它结果吻合很好。

结论 DSC 和 DMA 是能用于研究 PCL/PTHF 共聚物的熔融和结晶行为的很好的工具。用 DMA 测试的相转变温度与用 DSC 和 X-射线研究获得的结果吻合很好。结果显示结晶时的过冷随着 PTHF 含量的提高而提高,结晶时间因此延长。PTHF 似乎延迟了结晶。这对于特别应用例如医学应用可能是有意义的。PTHF 含量也影响模量,这意味着材料的机械性能可以依赖于应用调节到最优化。

这些结果可用来获得有关 PCL/PTHF 共聚物分子特性的较多信息。

stiffness, as the increase of the modulus shows.

In the second diagram, the shear moduli (G') of the three samples are compared during melting and crystallization. The modulus in the solid state decreases with increasing content of PTHF. The modulus seems to be low. It is in fact adjusted to this value so that the material is still stiff enough for the application, but on the other hand is not too brittle and does not break on mechanical shock.

The level of PTHF also determines the temperature and time of crystallization. With increasing PTHF content, the crystallization shifts to lower temperatures, i. e. isothermal crystallization would take longer.

The results are in good agreement with other results obtained by time-resolved small (SAXS) and wide-angle (WAXS) X-ray scattering analysis in the same temperature range.

Conclusions DSC and DMA are complimentary tools that can be used to investigate the melting and crystallization behavior of PCL/PTHF copolymers. The phase transition temperatures measured by DMA agree well with results obtained by DSC and X-ray investigations. The results show that super-cooling during crystallization increases with increasing PTHF content and that crystallization time therefore increases. PTHF seems to delay crystallization. This could be of interest for specific applications such as in medical use. The PTHF content also influences the modulus. This means that the mechanical properties of the material can be adjusted to optimum values depending on the application.

The results can be used to gain more information on the molecular properties of PCL/PTHF copolymers.

7.10 热塑性塑料及其产品的进一步测试
Further Measurements of Thermoplastics and Their Products

聚丙烯/聚乙烯共聚物的定性检查 PP/PE, Copolymer Identity Check

样品 **Sample**	球状颗粒 Stamylan P46M10 注射成型部件,据称是 PP/PE 共聚物 Stamylan P46M10 as pellet, injection-molded part, said to be a PP/PE copolymer	
条件 **Conditions**	测试仪器:DSC 坩锅:40µl 标准铝坩锅,钻孔盖 样品制备: 从球状颗粒/部件上切下圆片 DSC 测试: 以 10 K/min 从 30℃加热到 300℃ 气氛:空气,50 cm³/min	**Measuring cell**:DSC **Pan**:Aluminum standard 40µl, pierced lid **Sample preparation**: Disk cut from pellet/part **DSC measurement**: Heating from 30℃ to 300℃ at 10 K/min **Atmosphere**:Air, 50 cm³/min

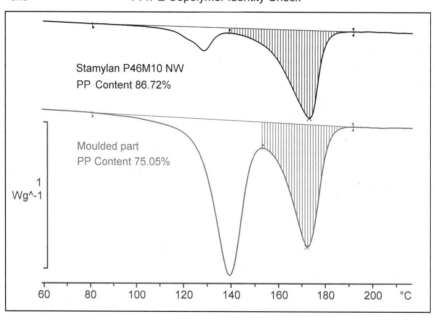

计算 **Evaluation**	第一个峰,PE First peak, PE	Stamylan P46M10 Stamylan P46M10	成型部件 Molded part
	熔融热,J/g Heat of fusion in J/g	14	71
	峰温,℃ Peak temperature in ℃	127.4	138.6

第二个峰，PE Second peak, PE	Stamylan P46M10 Stamylan P46M10	成型部件 Molded part
熔融热，J/g Heat of fusion in J/g	61	52.5
PP 含量，% PP content in %	87	75
峰温，℃ Peak temperature in ℃	171.4	171.2

解释 每条曲线显示两个熔融峰，一个为 PE 的，另一个为 PP 的。基于假定的 PP 熔融热 70J/g（对应于结晶度 34%）可以估计它们的相对量。

结论 两种聚合物是明显不同的，即成型部件不是由 Stamylan P46M10 制成的。
DSC 不仅是定性的，而且可进行共聚物的半定量分析。

Interpretation Each curve displays two melting peaks, one for PE and the other for PP. An estimate of their relative amounts is possible based on an assumed heat of fusion of 70 J/g for PP (corresponding to a degree of crystallinity of 34%).

Conclusions The two polymers are clearly different, i.e. the molded part is not made of Stamylan P46M10.

DSC allows not only qualitative, but also semiquantitative analysis of copolymers to be performed.

聚醋酸乙烯的玻璃化转变温度和增塑剂含量
PVAC, Glass Transition Temperature and Plasticizer Content

样品	含 0 到 12.5% 增塑剂的聚醋酸乙烯薄膜		
Sample	Polyvinylacetate films with 0 to 12.5% plasticizer		
条件	测试仪器：DSC	**Measuring cell**：DSC
Conditions	坩锅：40 μl 标准铝坩锅，密封	**Pan**：Aluminum standard 40 μl, hermetically sealed
	样品制备：	**Sample preparation**：Punched disks of approx. 5 mm diameter
	冲出直径约 5 mm 的圆片	
	DSC 测试：	**DSC measurement**：
	以 10K/min 从 25℃ 加热到 100℃ 以消除其热历史。不受控快速冷却到 −100℃。	Heating from 25℃ to 100℃ at 10 K/min to eliminate thermal history. Uncontrolled fast cooling to −100℃.
	实际测试以 10K/min 从 −100℃ 加热到 80℃	Actual measurement from −100℃ to 80℃ at 10 K/min
	气氛：氮气，50 cm³/min	**Atmosphere**：Nitrogen, 50 cm³/min

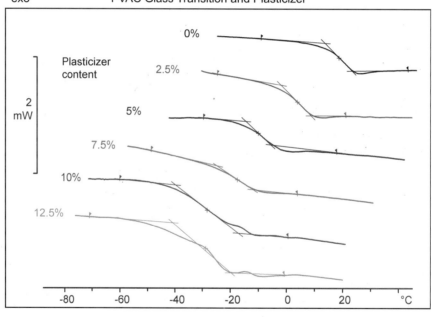

计算 表格总结了制造商给的增塑剂含量和 T_g 中点。中点是一半材料已经经历了玻璃化转变的样品温度。

图是由表格中的数值绘制的，显示了增塑剂含量对聚醋酸乙烯（PVAC）样品的玻璃化转变温度的影响。较高的增塑剂浓度造成玻璃化转变温度移向较低温度。

Evaluation The table summarizes the plasticizer contents as given by the manufacturer and the T_g midpoints. The midpoint is the sample temperature at which half of the material has undergone the glass transition.

The graph is plotted from the values in the table and shows the effect of the plasticizer content on the glass transition temperature of a sample of polyvinyl acetate (PVAC). Higher concentrations of plasticizer cause the glass transition temperature to shift to lower temperatures.

样品质量,mg Sample mass in mg	增塑剂含量 α % Plasticizer Content α in %	测试的 T_g,℃ T_g, measured in ℃
9.285	0	18.3
9.877	2.5	2.8
8.716	5.0	−10.2
9.410	7.5	−18.7
14.239	10.0	−29.1
13.795	12.5	−31.3

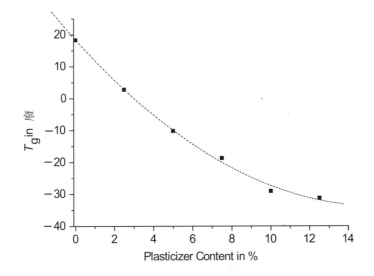

解释　为了清晰,DSC 曲线仅显示玻璃化转变发生的区域。正如预期的,增塑剂的存在降低了 T_g。随着增塑剂含量的增加玻璃化转变变宽。

没有增塑剂的 PVAC 的 T_g 测试值 18.3℃与通常报道的值 30℃不符。众所周知 PVAC 的 T_g 强烈依赖于水分含量。1%的水分含量降低 T_g 约 7K。因此水分是有效的增塑剂,但它是挥发性的。所以对一些材料,从空气中吸收的水可以起增塑剂的作用。源自材料制造或加工的溶剂残留也可能作用如(不受欢迎的)增塑剂。

Interpretation　For clarity, the DSC curves show only the regions in which the glass transition occurs. As expected, the presence of the plasticizer lowers the T_g. The glass transitions become broader with increasing plasticizer content.

The measured value of T_g of the PVAC without plasticizer of 18.3 ℃ does not agree with the value usually reported of 30℃. It is known that the T_g of PVAC depends strongly on the moisture content. A moisture content of 1% lowers the T_g by about 7 K. Moisture is therefore an efficient plasticizer, but it is volatile. Hence, with some materials, it is possible for water, absorbed as moisture from the air, to act as a plasticizer. Solvent residues originating from the manufacture or processing of the material can also behave as (unwelcome) plasticizers.

结论　增塑剂对聚合物玻璃化转变的影响能够通过对不同增塑剂含量的样品进行若干简单的 DSC 测试来确定。要注意消除由不同热历史和水分的存在所产生的干扰。

Conclusions　The influence of plasticizers on the glass transition of a polymer can be determined by performing a few simple DSC measurements with samples of different plasticizer content. Care should be taken to eliminate interferences due to different thermal history and the presence of moisture.

聚对苯二甲酸丁二醇酯共混物上涂膜的粘着性
Adherence of Paint on A PBT Blend

样品	两个批次的含有约5%玻璃纤维的Valox VX 5005（PBT、PC和EPDM的共混物）球状颗粒。 第一批次：涂料薄膜的粘着性差 第二批次：粘着性好	
Sample	Two batches of Valox VX 5005 (blend of PBT, PC and EPDM) pellets containing about 5% glass fiber. First batch: poor adherence of paint film Second batch: good adherence	
条件	测试仪器：DSC	**Measuring cell**: DSC
Conditions	坩埚：40 μl 标准铝坩埚，钻孔盖	**Pan**: Aluminum standard 40 μl, pierced lid
	样品制备：	**Sample preparation**:
	从球状颗粒上切下圆片（第一批次6.7mg，第二批次14.4mg）	Disk cut from pellet (first batch 6.7 mg, second batch 14.4 mg)
	DSC测试：	**DSC measurement**:
	以10K/min从室温加热到300℃	Heating from room temperature to 300℃ at 10 K/min
	气氛：氮气，50 cm^3/min	**Atmosphere**: Nitrogen, 50 cm^3/min

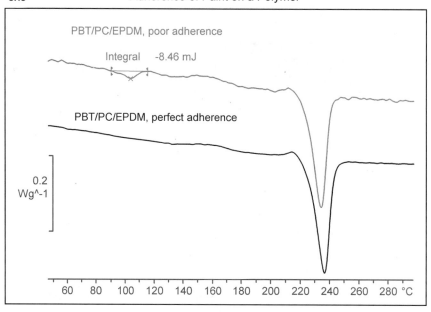

计算 滑爽添加剂的含量与约100℃处峰的熔融热成比例。第一批次的值为8 mJ，这对应于约1.2%的滑爽剂（假定该添加剂的熔融热为100J/g）。

Evaluation The content of slip additive is proportional to the heat of fusion of the peak around 100℃. The first batch gives a value of 8 mJ. This corresponds to approx. 1.2% slip agent (assuming a heat of fusion of the additive of 100 J/g).

解释 除了 234℃ 的 PBT 熔融峰，在 100℃ 左右有一个小峰。像这样的峰常常是由滑爽添加剂或成型润滑剂产生的。有些添加剂保留在注射成型件的表面，降低了涂膜的粘接性。在向供应商申诉后，发来了不含滑爽添加剂的第二批次材料，这批显示没有 100℃ 处的峰。

结论 如果注射成型部件必须涂膜，聚合物中添加剂的存在可能影响涂膜的粘接性。像蜡这样有结晶的添加剂可通过其 DSC 熔融峰来检测。

Interpretation Besides the PBT melting peak at 234 ℃, there is a small peak around 100 ℃. A peak like this is often caused by the melting of a slip additive or mold lubricant. Some of these additives remain on the surface of the injection-molded parts and lower the adherence of the paint film. After complaining to the supplier, a second batch of material was delivered without the slip additive. This showed no peak at 100℃.

Conclusions If injection-molded parts have to be painted, the presence of additives in the polymer may influence the adherence of the paint film. Additives such as waxes that undergo crystallization can be detected by their DSC fusion peak.

TMA 测试合成纤维　TMA Measurements on Synthetic Fibers

样品　纺织用纤维：PA66（95 dtex）、PET（25 dtex）和 Kevlar（50 dtex）
　　　　注：线和纤维的线密度以 dtex（g/10 km）来表示

Sample　Fibers for use as textiles：PA66 (95 dtex), PET (25 dtex), and Kevlar (50 dtex).
　　　　Note：The linear density of threads and fibers is expressed in dtex (g/10 km)

条件

Conditions

测试仪器：
TMA，由石英制造的纤维附件

样品制备：
两个距离约 13 mm 的 U 形铜夹将长约 20 mm 的纤维固定住。夹具能使纤维装在纤维附件中。

TMA 测试：
以 10 K/min 从 25℃加热到各自的结束温度

负载：
为了比较目的，对所有的纤维使用 0.1 mN/dtex 的拉伸应力。这小于大约 50 mN/dtex 即 500 N/mm² 拉伸强度的 1%。

气氛：静态空气

Measuring cell：
TMA with fiber accessory made of fused silica

Sample preparation：
Two U-shaped copper clamps about 13 mm apart are fixed to an approx. 20 mm length of fiber. The clamps allow the fiber to be mounted in the fiber accessory

TMA measurement：
Heating from 25℃ at 10 K/min to the individual end temperatures

Load：
For comparison purposes, a tensile stress of 0.1 mN/dtex is used for all fibers. This is less than 1% of the tensile strength of about 50 mN/dtex or 500 N/mm²

Atmosphere：Static air

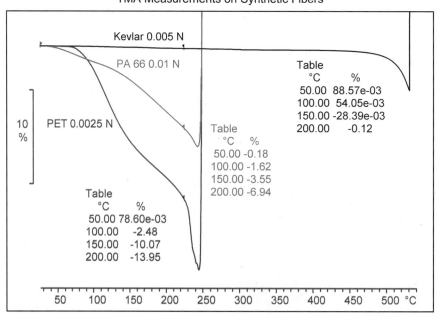

计算 作为例子,不同温度下相对于25℃时长度的长度变化用百分比进行计算(负号表示收缩):

Evaluation As examples, the change in length relative to that at 25℃ is calculated in percent at different temperatures (a negative sign indicates shrinkage):

温度,℃ Temperature(℃)	相对长度变化,% Relative length change in%		
	PET	PA66	Kevlar
50	0.08	−0.18	0.09
100	−2.48	−1.62	0.05
150	−10.07	−3.55	−0.03
200	−13.95	−6.94	−0.12

断裂温度 Temperature of rupture	245.0℃	244.4℃	530.8℃

解释 纤维的大分子在长度方向上是高度取向的。加热时,纤维由于干燥和在足够高的温度下形成无规线团而收缩。无规线团的无序度即熵比平行大分子高。这些效应通常过度补偿了膨胀。到达微晶熔融区域时,发生快速膨胀直到纤维最终断裂(例如:PET 在 245℃)。收缩的峰温可以用于材料的鉴定。

在 TMA 曲线上,收缩显示为向下的方向。

Interpretation The macromolecules of the fibers are strongly oriented in the longitudinal direction. On heating, the fibers shrink due to drying and-at sufficiently high temperatures-random coil formation. The disorder or entropy of random coils is higher than that of parallel macromolecules. These effects usually over-compensate the expansion. Toward the crystallite melting range, rapid expansion occurs until the fiber finally breaks (e.g. PET at 245℃). The peak temperature of shrinkage can be used for identification.

On a TMA curve, shrinking is displayed in the downward direction.

结论 加热时纤维的收缩或膨胀采用纤维夹具可用 TMA 容易地测试。即使非常小的形变也能检测,就如用芳香族纤维 Kevlar 所示的那样。TMA 曲线可用来鉴定纤维。

Conclusions The shrinkage or expansion of fibers on heating is easily measured in the TMA using the fiber accessory. Even very small deformations can be detected as is shown using the aramid fiber Kevlar. The TMA curve can be used to identify the fiber.

用 DMA 和 DSC 分析墨粉　Analysis of Toner Powder by DMA and DSC

样品

Sample

激光打印机中用的墨粉。该粉末由热塑性基本材料组成，其中混合了不同的添加剂如流动剂（石蜡）、颜料、UV 稳定剂和其他。分析的目的是测试特征玻璃化转变和熔融温度。

Toner powder as used in laser printers. The powder consists of a thermoplastic base material to which different additives such as flow agents (waxes), pigments, UV-stabilizers and others have been mixed. The purpose of the analysis was to measure the characteristic glass transition and the melting temperature.

条件

Conditions

测试仪器
DMA，剪切夹具样品支架
DSC

Measuring cell：
DMA with shear clamp sample holder，
DSC

样品制备：
粉末被压缩成圆柱体，厚 1.5 mm、直径 12 mm，装在 DMA 的剪切样品支架中。
DSC 用 40 μl 铝坩锅。

Sample preparation：
The powder was pressed to cylinders, 1.5 mm thick with a diameter of 12 mm, and mounted in the shear sample holder of the DMA.
In the DSC, 40 μl aluminum pans were used.

DMA 测试：
测试在 1、10、100 和 800Hz 下，2K/min。最大力振幅为 3N；最大位移振幅为 1μm；偏移控制为零。

DMA measurement：
The measurement was performed at 1, 10, 100 and 800 Hz at 2 K/min.
Maximum force amplitude 3 N; maximum displacement amplitude 1μm; offset control zero.

DSC 测试：
压成圆柱体的墨粉分别以 100、10 和 0.5 K/min 的速率加热。
气氛：静态空气

DSC measurement：
Toner powder pressed to cylinders was heated at rates of 100, 10 and 0.5 K/min respectively.
Atmosphere：Static air

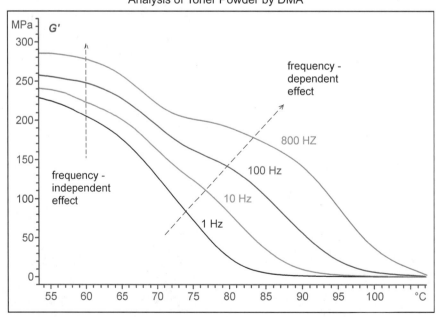

Analysis of Toner Powder by DMA

Interpretation The storage modulus DMA curves show frequency-dependent behavior. At 800 Hz, two steps are visible in the curve, whereas at 1 Hz only one main step can be seen. Since the first step at 800 Hz and the step at 1 Hz have about the same onset temperature, one must assume that a frequency-dependent and a frequency-independent effect overlap. Toners are complex mixtures, so one can conclude that the sample contains a component that melts at about 60 ℃ (melting is not frequency dependent). In contrast, the glass transition temperature of the thermoplastic base material depends on the applied frequency and shifts to higher temperatures at higher frequencies. The two effects are therefore separated at high shear frequency.

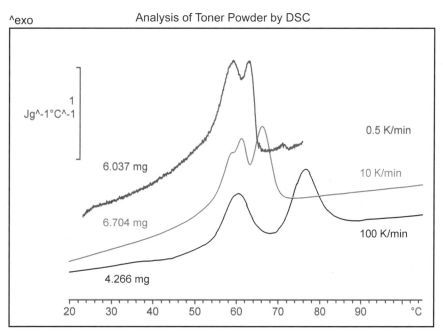

To confirm this interpretation, the same sample was also investigated by DSC. With this technique, the glass transition temperature depends strongly on the heating rate, whereas melting should be almost independent of the heating rate. This is shown by the specific heat capacity curves recorded at 0.5, 10 and 100 K/min The cp curves are shown here instead of the DSC curves because the size of the peaks in the heat flow curves would be too different to be sensibly compared. It can be clearly seen that the first endothermic peak, i.e. melting, is almost independent of the heating rate, whereas the second peak shifts to higher temperatures at higher heating rates. This peak corresponds to the enthalpy relaxation of the glass transition.

结论 重叠的熔融和玻璃化转变过程能够容易地用 DSC 和 DMA 分离。在 DSC 中，玻璃化转变依赖于加热速率的事实用于分离两个效应。对于 DMA，因为两个效应的频率依赖性是不同的，所以得到分离。在原则上，两种方法产生相同的结果。但是，DSC 对玻璃化转变的灵敏度比对熔融过程的灵敏度略低。不同的是，DMA 对于同样程度的分子运动的变化更灵敏。

Conclusions Overlapping melting and glass transition processes can be easily separated by both DSC and DMA. In DSC, the fact that the glass transition is dependent on the heating rate is used to separate the two effects. With DMA, separation is achieved because the frequency dependence of the two effects is different. In principle, both methods yield equivalent results. However, the sensitivity of DSC with regard to glass transitions is appreciably lower than the sensitivity towards melting processes. In contrast, DMA is sensitive towards changes in molecular mobility to the same extent.

文献 Literature

Thermal Analysis

Bernhard Wunderlich

Academic Press, San Diego, 1990, ISBN 0-12-765605-7

Calorimetry and Thermal Analysis of Polymers

V. B. F. Mathot

Hanser Publishers, New York, 1994, ISBN 3-446-17511-3

Thermal Analysis, Fundamentals and Applications to Polymer Science

T. Hatakeyama, F. X. Quinn

John Wiley & Sons, Chichester, 1994, ISBN 0-471-95103-X

Assignment of the Glass Transition

Rickley J. Seyler

ASTM STP 1249, Philadelphia, 1994, ASTM PCN ISBN 04-012490-50

Differential Scanning Calorimetry

G. W. H. Höhne, W. Hemminger, H.-J. Flammersheim

Springer-Verlag Berlin, 1996, ISBN 3-540-59012-9

Thermal Characterization of Polymeric Materials

Edith A. Turi (Ed.)

John Wiley & Sons, Academic Press, New York, 1996, Second Edition, ISBN 0-12-703783-7

Limitations of Test Methods for PLASTICS,

James S. Peraro (Ed.)

ASTM STP 1369, West Conshohocken, 2000, ISBN 0-8031-2850-9

UserCom, Information for users of METTLER TOLEDO thermal analysis systems,

published biannually by Mettler-Toledo GmbH, Analytical, Sonnenbergrstrasse 74, CH-8603 Schwerzenbach, Switzerland

Internet: www.mt.com/ta

图书在版编目(CIP)数据

热塑性聚合物：汉英对照/(瑞士)詹达利(Jandali,M. Z.)，
(瑞士)威德曼(Widmann,G.)著；陆立明，唐远旺，蔡艺译.
—上海：东华大学出版社，2008.6
ISBN 978—7—81111—390—7

Ⅰ.热… Ⅱ.①詹…②威…③陆…④唐…⑤蔡…
Ⅲ.热塑性—高聚物—汉、英 Ⅳ.O63

中国版本图书馆 CIP 数据核字(2008)第 083426 号

责任编辑　竺海娟
封面设计　蔡顺兴

热塑性聚合物

东华大学出版社出版
上海市延安西路1882号
邮政编码：200051　电话：(021)62193056
新华书店上海发行所发行　苏州望电印刷有限公司印刷
开本：889×1194　1/16　印张：10.5　字数：340千字
2008年7月第1版　2019年12月第2次印刷
ISBN 978—7—81111—390—7
定价：79.00元